I0486018

Cover photo: "Saddledome" (1983), Calgary, by Jan Bobrowski; a 136 m diameter prestressed stadium built using light-weight aggregate concrete.

Printed by CreateSpace, an Amazon.com Company.

Published 2016

Prestressed Concrete in Buildings

Niall MacAlevey

In memory of Dr. Jan Bobrowski
(1925-2014)

Preface

In design practice today it is usual to carry out the detailed design of members of conventional reinforced concrete, structural steelwork or prestressed concrete using computer software. Although much time is saved, this may have the unfortunate consequence that the much of the "feel" for the design is lost. This is particularly true in prestressed concrete, especially when indeterminate members are involved. Indeed the topic may be less well understood as many undergraduate courses neglect the teaching of the subject almost entirely. Thus the design of prestressed concrete members, in buildings particularly, is often left to specialist contractors. The focus of this book is to provide a conceptual understanding of the topic. The approach taken is design-based rather than a more rigorously analytical one. Hand-based methods are emphasized where relevant. Numerous worked examples are presented as well as extended examples on a pre-tensioned double tee floor and a post tensioned flat plate floor to illustrate the points made. The codes used are the European Code EC2 and the Concrete Society Technical Report TR43, while, for comparison, reference is sometimes made to other codes, e.g., the American Code ACI-318.

CONTENTS

CHAPTER 1: INTRODUCTION

1.1 PRESTRESSING AND PRESTRESSED CONCRETE

Prestressing is the deliberate pre-loading to introduce stresses that partially or fully counteract subsequent load-induced stresses.

Concrete is weak in tension, so prestressing the concrete in compression before loading is applied, counteracts tensions produced by load.

1.2 BRIEF HISTORY OF PRESTRESSED CONCRETE

1.2.1 P.H. Jackson (U.S.A.)

The first patent for prestressed concrete was taken out in 1886 by P.H. Jackson. He developed the notion of prestressing. However, his idea remained just that. There was a practical problem preventing the realization of his idea: most steels of that time were just not strong enough to overcome the **losses** due to creep and shrinkage of concrete. Thus, all early attempts to prestressed concrete failed but it was not understood why.

Illustration:

Suppose a 20 m long concrete member is to be prestressed. The steel is stressed, causing it to lengthen say **100 mm**, and then anchored to the concrete. The concrete is now compressed and shortens a small amount due to shrinkage as well as creep (we'll ignore elastic deformation of the concrete as it is small). Say this shrinkage and creep is **16 mm**, then the remaining steel extension is **84 mm**. Thus 16 mm worth of extension is "lost", i.e. $E\varepsilon$ = 210,000x16/20,000 = 168 N/mm^2. This is a large fraction of the yield strength of mild steel, the only steel which was cheap enough to use in construction at the time.

1.2.2   Eugene Freyssinet (France)

In 1927 he completed three long span arch bridges in conventional RC using very little steel. He observed their deformation (in one case the sagging of the arches was so large that it threatened to cause the collapse of one of the bridges (Billington 2004)) and knew it was due to shrinkage and creep of concrete. Thus he knew why previous attempts at prestressed concrete had failed.

In the 1920's high strength steel became readily available, and our understanding of shrinkage and creep improved as a result of research.

So by 1928 Freyssinet was able to solve the practical problem preventing PSC by using **high strength steel** to overcome long-term losses from shrinkage and creep of concrete. In addition, **good quality concrete** was used to reduce the magnitude of the creep.

In 1946 first major prestressed concrete structure was completed: bridge over the river Marne at Luzancy, France. It was a precast two-hinged arch spanning almost 60 m with access for jacks at hinges to allow later adjustment of profile.

Freyssinet's original idea was <u>full</u> Prestressing (ensured that concrete never went into tension).

1.2.3   Paul W. Abeles (Austria/U.K.)

In the 1940s developed the concept of <u>partial</u> prestressing (design such that some tension in the concrete was allowed). He designed many bridges for British Rail using this concept.

1.2.4   T.Y. Lin (China/U.S.A.)

In 1960s he developed the "Load Balancing" concept, which was an important tool for design; especially useful for the design of indeterminate structures.

1.3 Tendons

The steel doing the compressing of the concrete is usually known simply as the "tendon" but it may be any of the following:

1) <u>wire:</u> high-strength small diameter (3 mm-7 mm) rod;
2) <u>bar:</u> large diameter (25 mm-35 mm) rod;
3) <u>strand:</u> group of wires, commonly 7 (or 19 wires), in which six wires are spun around one central (straight) wire forming a helix;
4) <u>cable:</u> group of strands. (In the bonded post-tensioning system strands are often grouped within a single duct.)

1.4 Pre-tensioning and Post-tensioning

The terms "Pre-tensioning" and "Post-tensioning" strictly apply to the **steel** only.

1.4.1 Pre-tensioning

Here the tendons are tensioned before the concrete is placed. The procedure is commonly used in precasting plants. The steel is placed and then stressed using two concrete abutment placed maybe 100 m apart. Concrete is then cast around these tensioned tendons. The tendons are later cut, transferring the force to the concrete. The members are known as pre-tensioned members. Naturally, the tendon profile is usually straight but it can be harped (at greater expense). The permanent connection between concrete and steel is by bond.

1.4.2   Post-tensioning

In this case, the tendons are tensioned after concrete has hardened. A duct made of galvanized steel ensures the tendon can move reasonably freely when stressed. Almost any profile can be used, but as will be seen, the usual profile is parabolic.

1.5   TYPES OF PRESTRESSED MEMBER

A *fully prestressed member* has sufficient prestressing so that tensile stresses, for normal service loads, are eliminated. The member is said to be a *zero tension* member.

A *partially prestressed member* has less prestressing. Limited tensile stresses in the member under service loads are allowed. The member is said to be *uncracked* if the level of tension is low. Otherwise it is *cracked*.

1.6   DESIGN GUIDANCE

<u>1)</u>   BSI EN 1992-1-1: 2008, Eurocode 2 - Part 1-1: *Design of concrete structures - General Rules and Rules for Buildings*. The code covers beams and one-way slabs.

<u>2)</u>   Technical Report TR43, *"Post-tensioned Concrete Floors-Design Handbook"*, 2nd edition, 2005, UK Concrete Society. One and Two-way slabs. Latest in series of reports. Considered to be "the code" for PSC floors

1.7   ADVANTAGES AND DISADVANTAGES

1.7.1   Advantages

- **Deflection lower.**

Upward Load due to prestressing means less deflection under service load.

- **Greater stiffness and durability.**

Less cracking also means greater stiffness as more cross section is effective. Fewer cracks means more durable construction.

- **Reduction in weight/cost.**

Thinner and shallower sections possible (usually about 70% of the equivalent depth of RC can be used). Main saving in cost is due to the reduction in concrete.

Weight saving means easier lifting of precast elements, smaller loads to foundations, and smaller P-delta effects in multi-storey buildings.

Saving in height and volume of building means reduced cladding and air-con.

- **Reduction in number of expansion joints.**

Concrete is pre-compressed so can accept larger tensions resulting from restraint.

- **Less congestion; less steel to be placed.**

Prestressing enables higher strength steel to be used and so reduces steel congestion.

- **Faster Construction Time.**

Less steel to fix and concrete to pour. Less propping required. Floor slabs can become self-supporting a few days after the concrete is poured improving formwork turnaround.

- **Complete Rebound.**

Once live load is removed (provided load less than about 70% ultimate); vertical cracks close.

- **Progressive Collapse Resistance.**

Better, especially if tendons pass through column cage.

- **Under-strength member can often be found early.**

If the concrete and steel perform satisfactorily during prestressing, they will most likely perform well under service loading during their entire service life.

- **Especially economic if live load is low.**

Tendons provide an upward force so counteracting some or all of the dead load. Thus section needs to resist the unbalanced load only (see later).

- **Connections easier.**

Clamping force provided by post-tensioning means can often avoid requiring additional steel bars between joined elements, e.g. post-tensioning of glued segmental construction.

1.7.2   Disadvantages

- **More care/supervision is required.**

In design, construction, erection, etc.

- **Prestressing operation can be dangerous.**

Prestressing forces are often hundreds of tonnes. Clearly one should not stand behind a prestressing jack during stressing, or above a deflected pre-tensioned strand.

- **High temperature sensitivity.**

Because of its heat-treatment, prestressing strand is more sensitive to high temperature (e.g., due to fire).

- **Corrosion.**

Less steel generally and smaller diameter steel: high surface area/volume ratio, so prestressed members more susceptible. Also the possibility of stress corrosion is increased (as chemical bonds are stretched and more prone to attack).

- **Lack of flexibility (?)**

There is a common perception that future hole drilling is more difficult. Perhaps somewhat true if the unbonded system is used, as there is a risk of the tendon effect being lost over a long length when the tendon is cut. However, it is certainly not true if the bonded system used. The grout allows the tendon effect to be re-established either side of a cut. Actually the quantity of steel is less and that makes drilling easier.

- **Vibration of floors**

Floors are thinner than conventional RC so the natural frequency is lower. In addition. There are fewer cracks so less damping than RC.both of these features can lead to floor vibrations being a problem in PSC.

- **Differential camber in thin precast elements**

Affects thin elements most. Can be worsened by the unintentional use of aggegrates of different moduli and the tolerance in placing the tendon (typically 5-6 mm).

1.8   DESIGN OF PRESTRESSED CONCRETE

The member must be satisfactory throughout its life:

Design Check 1: "Transfer", i.e., immediately on application of prestressing.

Design Check 2: "Service", i.e., during normal service of structure.

Design Check 3: "Ultimate", i.e., there must be sufficient factor of safety against failure due to overload.

1.9 SOURCES OF LOSSES IN PRESTRESSED CONCRETE

Immediate losses: (during jacking or anchoring)

1. Friction between tendon and duct during stressing;
2. Anchorage slip: wedges bite into tendon;
3. Elastic shortening of concrete.

Long-term losses: (biggest in early years)

1. Creep of concrete (i.e. extra deformation in response to sustained stress);
2. Shrinkage of concrete;
3. Steel stress relaxation. (Just as sustained load on concrete leads to creep, high stresses sustained on steel leads it to creep too).

1.9.1 Prestress Force vs. Time

The following diagram indicates (Fig. 1.1) the changes to the applied force over time.

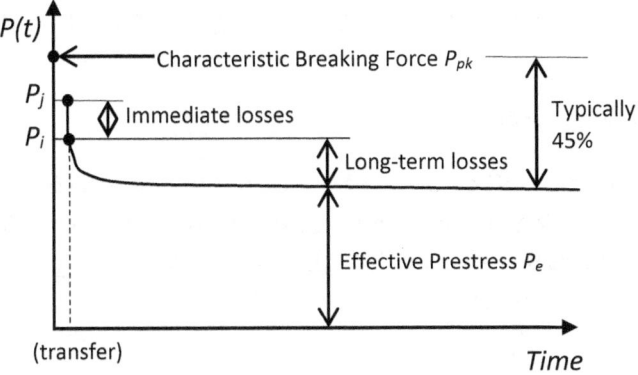

Fig. 1.1: Variation of prestress force over time

Although many books spend a large amount of time on methods to estimate losses (see e.g. Nilson (1997 and 1978) or Naaman (1982)), in building design such precision in estimating losses is rarely justified. The main reason a lump-sum value of losses can be simply guessed is the actual value of P_e does not have much effect on ultimate moment capacity so not on safety.

1.9.2   Typical values:

Jacking Force, $P_j \sim 0.75 P_{pk}$; Initial force, $P_i \sim 0.9 P_j$

(i.e., 10% force lost due to <u>immediate</u> losses)

Effective Force, $P_e \sim 0.55 P_{pk}$

(i.e., 25% force lost due to <u>all</u> losses).

12

1.10 CASE STUDY: EXCHANGE TOWER, LONDON, UK

Bennett (1991)

This is an office block of 17/18 floors. The use of Post Tensioning for the floors allowed an additional floor. The grid was 9 m x 7.5 m. The options considered are given in Table 1.1 below.

Table 1.1. Exchange Tower

Option	Floor depth (mm)	Total depth over 17 floors (m)	Cost index (frame only)
Steel			
9 m primary beams & 7.5 m secondary beams & 130 metal deck	410	6.97	1.35
7.5 m primary beams & 9 m secondary beams 130 metal deck	460	**7.82**	1.25
Concrete			
7.5 m PT beam & 9 m PT Ribbed Slab	250	**4.25**	1.0
PT flat slab with drops	250	4.25	1.05
7.5 m RC beam & 9 m RC Ribbed Slab	425	7.22	1.0

The difference in the post-tensioned option and the structural steel option was 3.57 m (7.82 − 4.25). This indicates that indirect cost savings of prestressed concrete floors can be significant.

1.11 BONDED AND UNBONDED TENDONS

These terms refer to the bond between concrete and post-tensioned tendons. The tendon must be relatively free to move when stressed.

Bonded: once stressing is complete, cement-based grout is injected into the duct under pressure.

Unbonded: individual tendons coated in anti-corrosion grease and wrapped in a plastic sheath to ensure there is no bond with the wet concrete.

Which to use is determined by the following:

1. Costs of labour vs. materials;
2. Design codes;
3. Speed of construction;
4. Relative costs of rebar vs. prestressing steel;
5. Cost of grouting;
6. Engineers/Clients preference (concern over corrosion or future modifications for example).

1.11.1 Unbonded Tendons

- Maximize eccentricity and are easy to handle and place.
- Grease acts as lubricant and results in low friction losses.
- The anchorage for the tendons is relatively small (each strand is individually anchored) so can easily be placed within columns, thus improving progressive collapse resistance.
- No grouting.
- The tendon rarely yields so more unstressed reinforcement required.
- Poor crack control, so add conventional rebar.

1.11.2 Bonded Tendons

- Results in a higher ultimate capacity as the tendon's yield strength is usually achieved.

- The tendon is better at controlling the widths of cracks that form. Thus less rebar must be added.

- Some durability concerns, e.g., as grouting may not be thorough, but bonded tendons in buildings have a better durability record than unbonded.

However, today in many countries in Europe and Asia the bonded system is used almost exclusively. The use of the unbonded system is common in the USA.

Chapter 2: Flexural Stresses at Service and Transfer

2.1 Behaviour of a Prestressed Beam

Consider a prestressed beam under increasing load. The load and mid-span deflection are measured. The following graph shows what happens as the load is increased (Fig. 2.1). There are three distinct stages. When loads are low the beam is uncracked and so at its stiffest ('elastically rigid' response). Once visible cracking appears the beam loses stiffness ('elastically flexible'). Deflections are still recoverable if the load is removed (assuming of course no debris prevents a crack from closing). Once the tendon yields the 'plastic' response is obtained. The beam is still stable under load. Complete loss of load carrying capacity, and so instability, occurs when the concrete in the compression zone crushes.

In many cases the service load is such that the beam is often totally in compression under frequent load, defined by EC2 to be 1.0DL+0.5LL for a residential or office structure. Thus our beam is probably in the 'rigid' part of the graph at service.

Clearly the beam also behaves in a 'rigid' way at the transfer stage too.

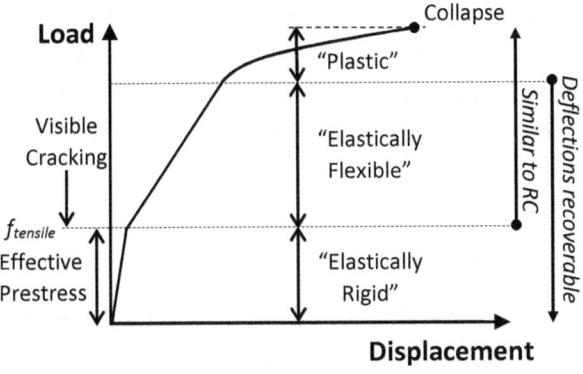

Fig. 2.1: Characteristics of typical prestressed beam

2.1.1   Flexural Stresses

We saw in the last section that the member usually behaves elastically at **service** and **transfer**. Using linear elastic theory flexural stresses can be checked.

Note that transfer, serviceability, and ultimate are independent. So for example, ensuring flexural stresses at transfer are below limits tells us nothing about service or ultimate conditions.

2.1.2   Checking Stresses-Why?

•   Service Stresses:

We must check these because if the tensile stresses are too high there is risk of local damage (cracking). In

18

addition, high sustained compressive stresses may eventually damage the concrete. (High stress > $0.85f_{ck}$ will eventually cause failure of the concrete. The phenomenon is known as Static Fatigue.)

- Transfer Stresses:

As the concrete is young local damage is more likely. Also, creep losses are greater if compression stresses are high.

2.1.4 Understanding Stress Directions

Consider a cut through end of a simply supported prestressed beam (Fig. 2.2). Ignore the self-weight of beam. Thus $R = 0$.

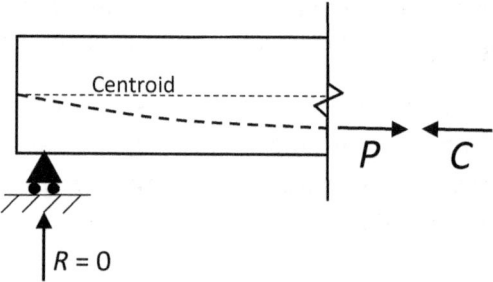

Fig. 2.2: End of simply supported beam

For rotational equilibrium C must equal P and be on same line. The C causes prestressing of the concrete. Thus there is an imaginary line of compression force in concrete which follows tendon line exactly. This is known as the "C-line".

Use the C-line to visualize what happens when the beam is prestressed. In the case shown we have Hogging Bending Moment imposed by prestress, i.e., the beam bends upwards (Fig. 2.3).

Notice that above structure is statically determinate (a simple beam), so prestressing itself causes no reactions, ($R = 0$).

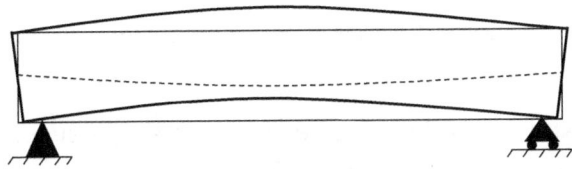

Fig. 2.3: Beam bends upwards ("hogs") under effect of prestress alone.

2.1.5   Stresses in Prestressed Concrete

Super-position is used, i.e., components of stress simply added together. Typically there are three separate items:

- Axial effect of prestressing (P/A)
- Bending effect of prestressing (Pe/Z)
- Effect of external applied loads (M/Z)

where A is concrete area, e is tendon eccentricity from centroid, and Z is section modulus (I/y_{max}).

2.1.5.1 Sign Convention

"1" = bottom of beam; "2" = top of beam;

+ ve stress in concrete = compression stress;

e = eccentricity is considered positive when tendon is *below* c.g.c.

M = considered positive when moment due to applied load is *sagging*.

The stresses at the extreme fibres of a simply supported beam are shown below (Fig. 2.4).

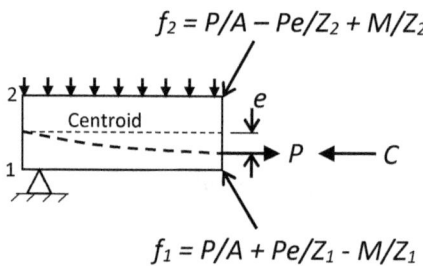

$$f_2 = P/A - Pe/Z_2 + M/Z_2$$

$$f_1 = P/A + Pe/Z_1 - M/Z_1$$

Fig. 2.4: Stresses at extreme fibres.

2.2   EXAMPLES OF CALCULATION OF STRESSES

2.2.1   Example 1

A precast (pre-tensioned) beam spans 10m simply supported. If the effective prestress P_e = 1000 kN and imposed UDL is 10 kN/m, calculate service stresses under characteristic loads at midspan and at top of beam near support.

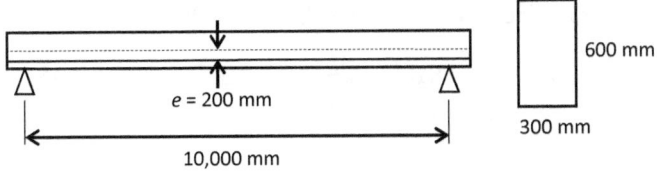

Total UDL w = 10 kN/m + Self-weight of beam = 10 + 0.3x0.6x25 = 14.5 kN/m

Applied moment at midspan $M = wL^2/8 = 14.5 \times 10^2/8$ = 181 kNm

Section modulus $Z = \frac{1}{6} bd^2 = 300 \times 600^2/6 = 18 \times 10^6$ mm^3

Solution:

• Midspan Stress at **Top** of beam:

= Axial effect of prestressing + Bending effect of prestressing + effect of applied load

22

$= P_e/A - P_e e/Z + M/Z$

$= 1000 \times 10^3/600 \times 300 - 1000 \times 10^3 \times 200/18 \times 10^6 +$

$\quad 181 \times 10^6/18 \times 10^6 = 5.6 - 11.1 + 10.1$

$= 4.6 \text{ N/mm}^2$ (compression)

- Midspan Stress at **Bottom** of beam:

= Axial effect of prestressing + Bending effect of prestressing + effect of applied load

$= P_e/A + P_e e/Z - M/Z$

$= 1000 \times 10^3/600 \times 300 + 1000 \times 10^3 \times 200/18 \times 10^6 -$

$\quad 181 \times 10^6/18 \times 10^6 = 5.6 + 11.1 - 10.1 = 6.6 \text{ N/mm}^2$

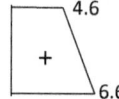

- Support stresses at **top** of beam

= Axial effect of prestressing + Bending effect of prestressing + effect of applied load

$= P_e/A - P_e e/Z + M/Z$

$= 1000 \times 10^3/600 \times 300 - 1000 \times 10^3 \times 200/18 \times 10^6 + 0 = 5.6 - 11.1 + 0$

$= -5.5 \text{ N/mm}^2$ (i.e., tension)

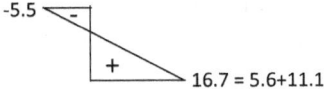

-5.5

16.7 = 5.6+11.1

Notice: there is a large tension at the support if we have a straight tendon. Can we do anything about this?

2.2.2   Example 2

A prestressed beam spans 11 m. It has a rectangular cross section 600 mm x 200 mm and is pre-stressed with a harped tendon and an effective prestress P_e = 500 kN.

Imposed load at service is 7 kN/m, and self-weight = 3 kN/m.

Given: A_c = 120×10^3 mm^2; e_{end} = 0 ; e_{max} = 200 mm;

Z_1 = Z_2 =12×10^6 mm^3.

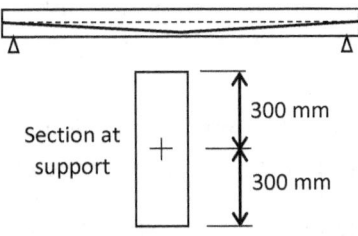

Section at support

300 mm

300 mm

Calculate the bottom fibre stress at the section $0.4L$ from the support, and the top fibre stress at the support under characteristic loading.

Solution:

- Stress at $0.4L$:

$M_{0.4L}$ $= 0.12wL^2 = 0.12(3+7)11^2 = 145.2$ kNm

$e_{0.4L}$ $= 0.8 (200) = 160$ mm

f_1 $= P_e/A + P_e e_{0.4L}/Z_1 - M_{0.4L}/Z_1$

   $= 500/120 + 500\times10^3\times160/12\times10^6 - 145.2/12$

   $= 4.16 + 6.67 - 12.1$

   $= -1.27$ N/mm^2 (i.e., tension)

(Note: at midspan, f_1 is -0.1 N/mm^2)

- Stress at top of beam at Support:

f_2 $= P_e/A = 500\times10^3/120\times10^3$

   $= +4.16$ N/mm^2 (i.e., compression)

Notice the "cure" for the excessive tension at the top of the beam near the support is to reduce the eccentricity.

- Post-tensioning: simply ensure the tendon is bent up near the support;

- Pre-tensioning: it is more convenient to de-bond some tendons so reducing P near the support. Wrapping in a plastic sleeve to prevent bond is a common way of de-bonding.

2.3 CHECKING STRESSES AT TRANSFER AND SERVICE

2.3.1 Transfer Check:

Prestressing force and gravity moment: $P_i + M_i$

Assume highest value of P likely as compression stresses are likely to be critical.

At **Transfer** (i.e., after short-term losses):

$f_{1(bot)} = P_i/A_c + P_i e/Z_1 - M_0/Z_1$

$f_{2(top)} = P_i/A_c - P_i e/Z_2 + M_0/Z_2$

2.3.2 Service Check:

Prestressing force and gravity moment: $P_e + M_s$

Assume lowest value of P likely as tension stresses are likely to be critical.

At **Service** load condition (after all losses):

$$f_{1(bot)} = P_e/A_c + P_e\,e/Z_1$$

$$- M_0/Z_1 - M_{DL}/Z_1 - M_{LL}/Z_1$$

$$f_{2(top)} = P_e/A_c - P_e\,e/Z_2$$

$$+ M_0/Z_2 + M_{DL}/Z_2 + M_{LL}/Z_2$$

where $M_i = M_o$ (i.e., s/wt); $M_s = M_o + M_{ADL} + M_{LL}$

2.4   STRESS LIMITS: ONE-WAY MEMBERS

2.4.1   Tensile

There are three types of member in EC2:

Type 1: Zero tension.

Type 2: Uncracked at service.

Type 3: Cracked at service.

The boundary between 2 and 3: If maximum tension under service load $\leq f_{ctm}$ (from Table 2.1) then section considered "uncracked". Otherwise it is "cracked".

Table 2.1: Concrete Properties (EC2)

Strength (MPa) and deformation (GPa) properties for concrete						
f_{ck}	C25	C30	C35	C40	C45	C50
$f_{ck,cube}$	30	37	45	50	55	60
f_{cm}	33	38	43	48	53	58
f_{ctm}	2.6	2.9	3.2	3.5	3.8	4.1
f_{ctk}	1.8	2.0	2.2	2.5	2.7	2.9
E_{cm}	31	33	34	35	36	37

As stated before EC2 also limits compression stresses to guard against static fatigue.

2.4.2   Service Loading: defined in EC2

1.   "Rare": i.e. Characteristic loads:
     (i.e. gravity load is $DL + LL$)
2.   "Frequent":
     (i.e. gravity load is $DL + \psi_1 LL$)
3.   "Quasi-permanent"
     (i.e. gravity load is $DL + \psi_2 LL$)

Where factors ψ_1 and ψ_2 are shown in the Table 2.2.

Table 2.2: factors for live load

Action	Frequent ψ_1	Quasi-permanent ψ_2
Residential area	0.5	0.3
Office area	0.5	0.3
Shopping area	0.7	0.6
Storage area	0.9	0.8
Roofs	0.0	0.0

2.4.3 Allowable Stresses in Concrete (MPa)

Under characteristic loads unless noted otherwise (EC2)

	Transfer	Service
Compression	$0.6f_{ck}^*$	
Characteristic Loads		$0.6f_{ck}$
Quasi-Permanent Loads		$0.45f_{ck}$
Tension		
Type 1 ("zero tension")	1.0	0
Type 2 ("uncracked")	f_{ctm}^*	f_{ctm}
Type 3 ("cracked")	f_{ctm}^*	see below

f_{ck}^* is characteristic compressive strength and f_{ctm}^* is mean tensile strength at time of transfer

2.4.4 Type 3 Allowable Stresses (MPa)

One-way slabs and beams (under **frequent** load) from TR43:

Tendons	Limiting Crack Width (mm)	In Tension
Bonded	0.1	$1.35\,f_{ctm}$
	0.2	$1.65\,f_{ctm}$
Unbonded	-	$1.35\,f_{ctm}$

Note: where additional unstressed reinforcement is placed close to faces stresses may be increased. For 1% => 4 MPa. Prorate up to $0.3f_{ck.}$

Analysis of section is based on the full value of second moment of area I. The allowable stress exceeds tensile strength. Thus there will almost certainly be cracks. Hence the stresses are 'hypothetical' (i.e. imaginary.)

2.5 "TYPES" OF CONSTRUCTION

The designation of type 1, 2 or 3 has no relation to <u>quality</u> of construction. Type 3 members are most suitable for buildings for the following reasons:

- More economical since less prestressing force P,
- The warning of failure greater (see below),
- There is less restraint cracking (since P/A lower),
- A larger amount of unprestressed reinforcement is used, so the fire resistance higher,
- The ability to absorb shocks (e.g., explosions) is greater (as section is more ductile).

Our visual cue that collapse is coming is large displacement/cracking. The figure below (Fig. 2.5) indicates that, for the same collapse load, more warning is obtained if a type 3 section is used rather than a type 1.

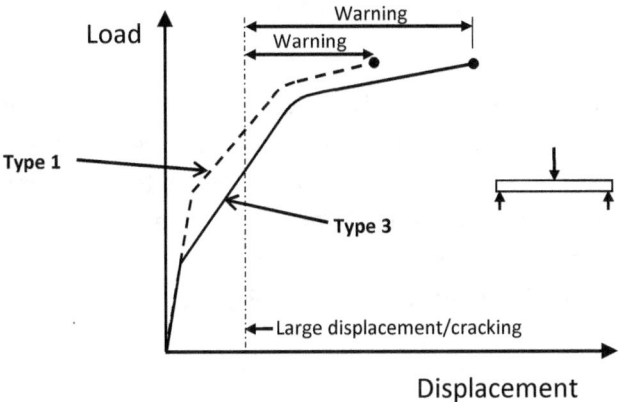

Fig. 2.5: Warning of failure

The main disadvantages of type 3 members compared to type 1 is that they have less fatigue resistance and are less durable (but of course still more durable than conventional RC). However, most prestressed concrete elements used for buildings is type 3 (except for pretensioned element which are usually type 2). Many road bridges are type 1 (appropriate for an exposed structure such as a bridge). Recall Freyssinet insisted on type 1, while Abeles advocated the use type 3. In fact, many road and rail bridges designed by Abeles in the early 20th century were type 3.

2.5.1   Cracked Section Analysis

As mentioned previously, we can ensure the stresses in a type 3 section are satisfactory by using the hypothetical stress limits in TR43. However, if we need to work out the crack widths or the deformation, we usually must analyze the section. The distribution of strain and stress in a type 3 section are shown in Fig. 2.6.

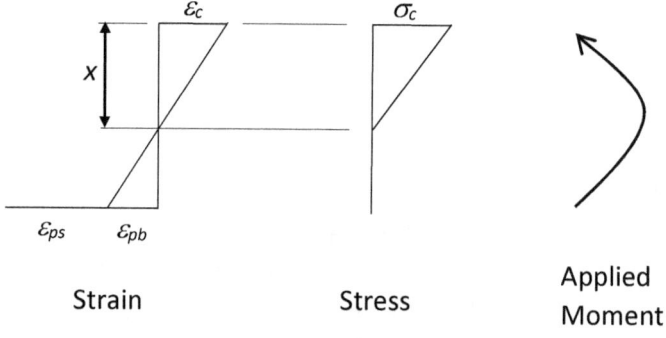

Strain   Stress   Applied Moment

Fig. 2.6: Analysis of cracked section

The procedure for section analysis is as follows:

(a) Choose a strain, ε_c

(b) Choose a neutral axis depth, x

(c) Determine the consequent concrete and steel stress from the material stress/strain curves, neglecting the concrete in tension below x.

(d) Check whether the sum of compression forces equals that of tension forces. If so, calculate moment of resistance.

(e) Repeat steps (a) to (d) until the moment of resistance equals the applied bending moment.

Clearly this procedure, involving the iteration of two variables, is best suited to a computer!

2.5.2   Crack width limit for cracked members

Normally, the crack width limit is 0.2 mm. (For liquid retaining structures the limit is 0.1 mm).

2.5.3   Decompression

If the environment is aggressive (e.g., seawater present) then all of the tendons should lie at least 25 mm within the compression zone (referred to as "decompression").

2.5.4   Comments on tensile limits

Suppose our trial solution passes the transfer and ultimate checks but fails the service check because the tensile stress limits are exceeded (narrowly). How concerned should we be? In my opinion *not very.* Here are my reasons:

1   Stresses are a poor guide to cracking;

For example, consider a pre-tensioned beam (f_{ck} = 55 N/mm^2), with a maximum tensile stress of 6 N/mm^2. The degree of cracking depends very much on distribution of steel:

- Bad (widely spaced steel and not placed close to surface): extensive visible cracking. Cracking hair-line only, but loss in second moment of area considerable (up to 70%),
- Good (narrowly spaced steel and placed close to surface): no visible cracking (only harmless microscopic cracking).

2   The elastic analysis itself may not be very accurate for indeterminate structures;

For a statically **indeterminate** beam the elastic solution is quite sensitive to the support conditions. For example consider a propped cantilever: we usually assume that the supports are completely rigid and then analyze the beam elastically. In fact, for a typical code-designed beam, if the roller support settles by only span/1000 (i.e., 10 mm for a span of 10 m) downwards relative to the fixed support, for a UDL, this leads to an increase in moment at the fixed support of about 15%. (Heyman 2008). See Fig. 2.7.

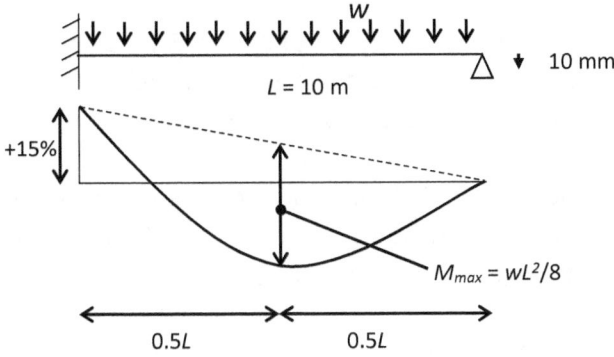

Fig. 2.7: Elastic solution very sensitive to boundary conditions.

The value of the static moment must be correct but we can't be sure of anything else. The point is we cannot be really sure our analysis reflects the actual stress. The chances are high that it does not.

3   Tensile strength is available only once and it may be reached inadvertently during construction. So the correct limit at service may in fact be zero;

4   The upper columns are usually absent when the prestress is applied to the floor. However, this is usually ignored when we analyze the structure.

5   Sustained load decreases tensile strength, by around 30% according to TR67 (2008). This is likely due to the fact the bonds are stretched and so are more vulnerable to environmental deterioration ;

6   The member rebounds once the LL is removed (as long as tendon does not approach yield). Thus it is unlikely that durability would be impaired;

7   The likelihood of the structure experiencing the full working LL is non great (around 10% in a typical building over a 50-year design life);

8   In conventional RC we never check service stresses.

CHAPTER 3: FLEXURAL STRENGTH

Member proportioning, determination of prestress force, and eccentricity are normally done at the serviceability limit state.

For this limit state partial safety factors for loading are taken to be 1.0 and stresses are limited to avoid local damage.

Analysis of ultimate strength of members is needed mainly to ensure adequate under factored load (over-loading). Thus it is a safety check.

3.1 SAFETY FACTORS

Load factors and materials factors according to Eurocode 2 are used.

i.e.,   Dead Load *(DL)* = 1.35 and

Imposed Live Load *(LL)* = 1.5

Where the partial safety factors are:

Concrete: γ_m = 1.5

Steel: γ_m = 1.15

The critical moment for the strength check for statically determinate members is the, **moment due to 1.35DL + 1.5LL** while that for statically indeterminate members is the **moment due to 1.35DL + 1.5LL + (worse of 1.1 and 0.9)M₂**

where M_2 is secondary moment (see later). For statically determinate structures $M_2 = 0$.

3.1.1 *Design* stress-strain curve for concrete in compression (EC 2)

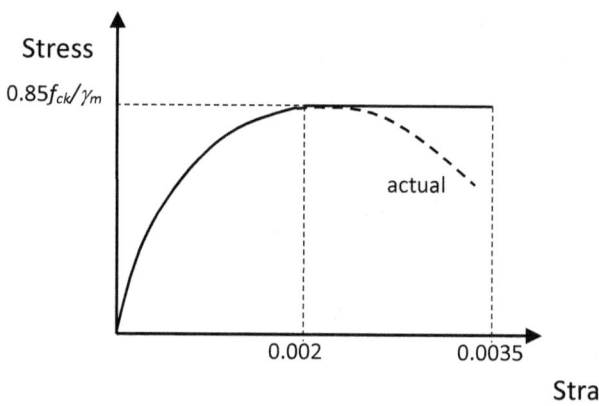

Because the actual state of stress near ultimate is triaxial compressive (see for example Kotsovos 2014) rather than simply uniaxial we can accept that the concrete behaves in the 'ductile' way shown.

3.1.2 *Actual* stress/strain curve for prestressing steel

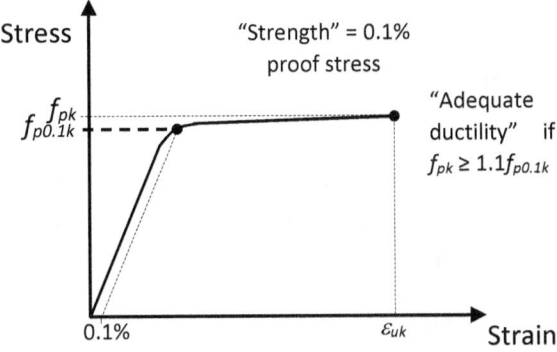

f_{pk} is the characteristic breaking strength. ε_{uk} is the characteristic strain corresponding to the peak stress. There is no distinct yield point for this type of steel. Preferably ε_{uk} should be at least 0.035 and then the allowable redistribution at ultimate is a maximum of 30%.

3.1.3 *Design* stress/strain curve for prestressing steel

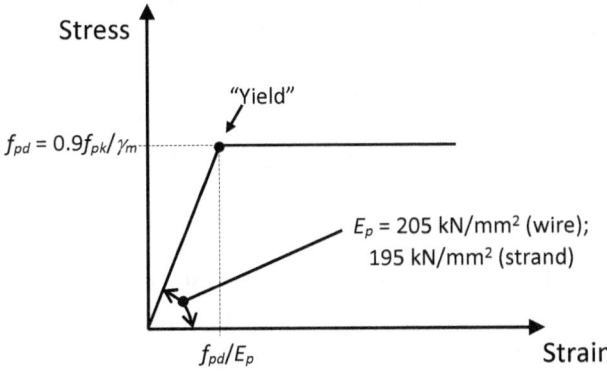

The difference between the modulus of wire and strand as latter has 6 non-straight wires in 7-wire strand.

3.1.4 Typical values for tendons:

Strand Type	Nominal tensile strength (MPa), f_{pk}	Cross-sectional area (mm²)	Characteristic maximum force (breaking force) (kN), F_{pk}
12.9 "super"	1860	100	186
15.7 "super"	1770	150	265
15.2 "drawn"	1820	165	300

3.1.5 Flexural Strength (Bonded Tendons)

The method of calculation used is quite similar to that used in the calculations for conventional RC design.

Similar assumptions are made, i.e.

- plane sections remain plane under loading;

- perfect bond exists between concrete and steel;

- concrete stress-strain diagram is shown in 3.1.1; limiting concrete strain ("crushing") 0.0035 corresponding to a stress of $0.57f_{ck}$;

- steel stress-strain curve as shown in 3.1.3;

- tensile strength of concrete ignored;

- rectangular stress block.

The only complication is that the prestressed steel is already **strained** before the member is loaded with gravity loads.

Thus at ultimate the stress and strain diagrams are as shown in Fig. 3.1. EC2 requires a factor of safety on the estimate of prestrain so $\varepsilon_{ps} = \gamma_p(P_e/A)$ where γ_p is a factor of safety and is 0.9; ε_{pb} = the additional strain due to bending; thus $\varepsilon_{pTOT} = \varepsilon_{ps} + \varepsilon_{pb}$ where ε_{pTOT} = total strain in steel.

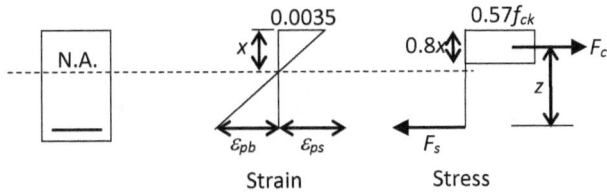

Fig. 3.1: Conditions at Ultimate Flexural Failure

Note we have made an engineering approximation to simplify the calculations: we have ignored pre-strain in concrete. This pre-strain, which strictly should be subtracted from ε_{cu} = 0.0035, is usually small. E.g. If P/A is 5 N/mm^2 and E of concrete is 33,000 N/mm^2 then the pre-strain in concrete is σ/E which is 5/33,000 = 0.00015 (i.e., 4% of 0.0035). It is thus reasonable that this is ignored.

3.1.5.1 Example 1

Find ultimate moment capacity of pre-tensioned beam section shown below. f_{ck} = 30 N/mm^2, and f_{pk} = 1800 N/mm^2. Wires have an effective prestress of 1080 N/mm^2.

Solution: Conditions at ultimate:

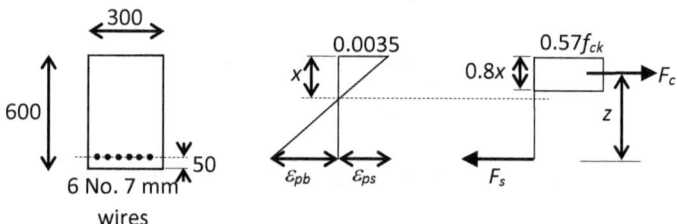

Stress/strain curve for wires:

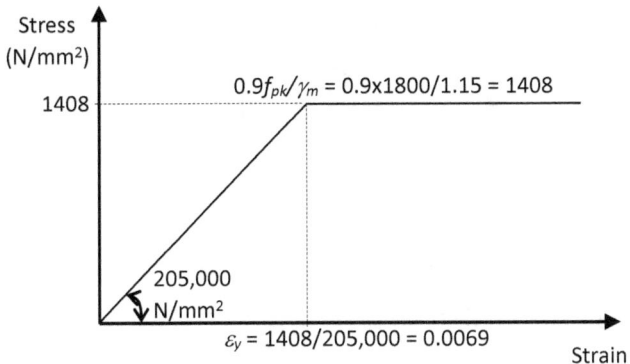

We first find neutral axis x

Let's assume prestressing wires have yielded.

Thus total *tensile* force F_s can be calculated:

$F_s = 6\, A_{wire}\, 0.9 f_{pk}/\gamma_m = 6\,(38.5)\,(0.9*1800/1.15) = 325$ kN

The total *compressive* force F_c in concrete is:

$F_c = 0.57\, f_{ck}\,(0.8x)\, b = 0.57\,(30)\,(0.8)\,(300)\, x$

Now, for equilibrium, $F_c = F_s$

$x = 325,000/(0.57*30*0.8*300) = 79$ mm

Now calculate strain in steel to see if it has yielded as assumed:

1. Prestrain $e_{ps} = \gamma_p f_{ps} / E_{ps} = 0.9x1080 / 205,000 = 0.0047$

2. Additional strain in steel due to bending is ε_{pb}

$$= (d-x)(0.0035)/x = (550 - 79)(0.0035)/79 = 0.021$$

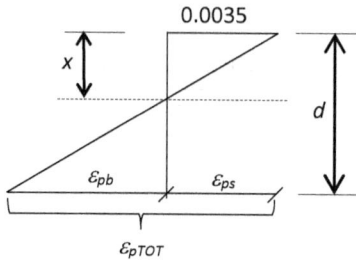

3. The total strain $\varepsilon_{pTOT} = \varepsilon_{ps} + \varepsilon_{pb}$

$= 0.0047 + 0.021 = 0.025 > \varepsilon_y = 0.0069$

=> Assumption correct, i.e., the prestressing steel has yielded.

Compute Ultimate Moment Capacity

Lever arm $z = d - 0.4\,x = 550 - 0.4\,(79) = 518$ mm

$$=> M_u = F_s \cdot z = 325\,(518)(10^{-3}) = \underline{168\ kNm}$$

Now suppose there was a mistake made during manufacture of these pre-tensioned beams and none of the tendons were stressed.

Thus $\varepsilon_{pTOT} = \varepsilon_{pb} = 0.021 > \varepsilon_y = 0.0069$

So the ultimate behaviour would not be affected. However, the service performance of this beam would certainly be impaired.

This leads to the following important observation:

A greater flexural moment of resistance of a section usually <u>cannot</u> be obtained merely by prestressing a beam.

Note: Tendon placement is governed by **ultimate** conditions: Thus the tendon is placed high over continuous supports and low in the span (i.e. approximately follows the BMD).

Use the <u>maximum eccentricity</u> available given cover requirements, so that the force required is a minimum (thus area of steel is a minimum).

For simplicity base the preliminary design on the <u>idealized</u> tendon profile. Detailed calculations can allow for the actual tendon profile.

CHAPTER 4: SHEAR STRENGTH

4.1 FAILURE OF CONCRETE MEMBERS

Usually the collapse of a beam is ultimately due to failure of the concrete. Concrete usually fails in a fairly brittle way. Steel usually fails is a ductile way. Brittle behaviour is unacceptable for beams as no warning (i.e. large deflections, cracking) of failure is given. Good design and detailing can make the behaviour acceptable.

4.1.1 TYPES OF FAILURE OF CONCRETE MEMBERS

1. *Ductile*: yielding of main reinforcing steel takes place <u>before</u> concrete crushes. For example under-reinforced flexural failure (a failure characterized by vertical cracks).

2. *Brittle*: concrete crushes <u>before</u> main steel fully yields. For example shear failure (a failure characterized by propagation of diagonal cracks), or over-reinforced flexural failure.

4.1.2 RC SIMPLY SUPPORTED BEAMS

Consider a series of identical rectangular RC beams, <u>without links</u>, loaded to failure by two equal point loads placed symmetrically about mid-span (see Fig. 4.1).

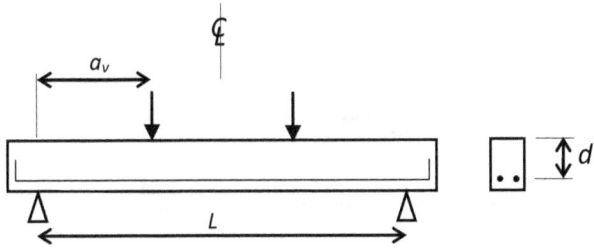

Fig. 4.1: Series of RC beams

The length a_v is called the "shear span". The loads are increased until the beam collapses. Often, the flexural moment M_{flex} is not reached and a brittle, non-flexural failure results, i.e. a "shear failure".

It usually happens because of the formation and propagation (widening and lengthening) of an inclined crack as the load is increased. The inclined crack eventually reduces the depth of the compression zone and results in failure of beam (i.e. the <u>concrete</u> fails).

Now if the total load at failure is P_{fail}, and the maximum moment sustained is $M_{fail} = P_{fail}a_v/2$.

The following is typically obtained from a series of these tests. As can be seen from Fig. 4.2, the behavior falls into four types (Kotsovos 1999):

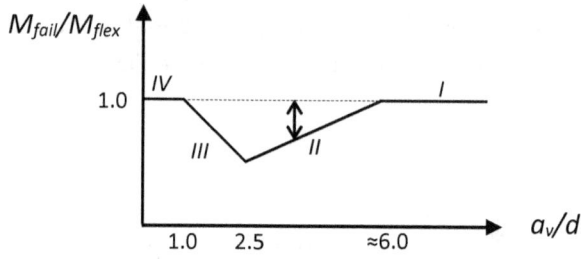

Fig. 4.2: Results of tests on RC beams

From the four zones in diagram we can identify four types of behaviour:

(1)   Type *I*: $a_v/d > \approx 6$: "Flexural Failure"

(2)   Type *II*: $2.5 > a_v/d < 6$: "Shear Failure"

(3)   Type *III*: $1 > a_v/d < 2.5$: "Shear Failure"

(4)   Type *IV*: $a_v/d < 1$: "Deep Beam Failure"

The purpose of adding shear reinforcement is to ensure that the flexural moment M_{flex} is reached. Excess links have no effect on the capacity. The degree that shear reinforcement is required is indicated by the arrow in Fig. 4.2. Clearly in some cases we do not really "need" any links at all (type *I* beams).

4.2   PRESTRESSED CONCRETE BEAMS

"...claims of basic differences between reinforced and prestressed concrete at the <u>ultimate limit state</u> can be classed as hardly more than superstitions."

Dr. Jan Bobrowski (1982).

Consider a beam with a straight tendon: prestressing is clearly beneficial at the service load level as the diagonal tension is reduced by the axial compression induced by the prestressing (see Fig. 4.3).

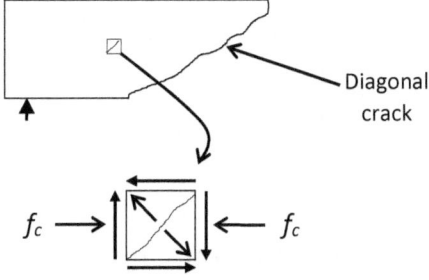

Fig. 4.3: Beneficial effect of pre-compression on shear

However, this benefit should really not be relied upon at the *ultimate* load level. <u>At ultimate</u> PSC with a straight tendon and RC behaviour is similar. However major codes of practice, including European and American codes, give different methods of assessment of shear strength for PSC members and an RC members.

4.2.1 PSC simply supported beams

Just as in the case of the RC beams above if we were to test a series of PSC beams with straight bonded tendons we would, in all likelihood, find this kind of diagram (see Fig. 4.4).

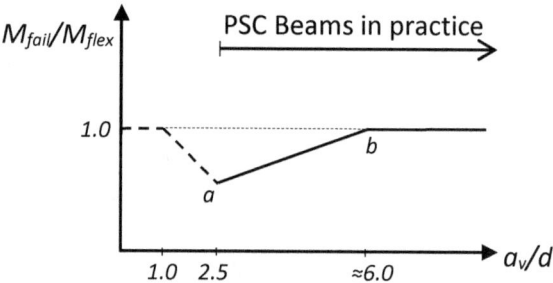

Fig. 4.4: likely behaviour of PSC beams

4.2.1.1   Illustration

Consider a simply supported prestressed member supporting *UDL*. Typically *L/h* about 22. Taking a_v as about $L/4$ (see below) and $d = 0.85h$, this means that a_v/d is 22/3.4 = 6.5.

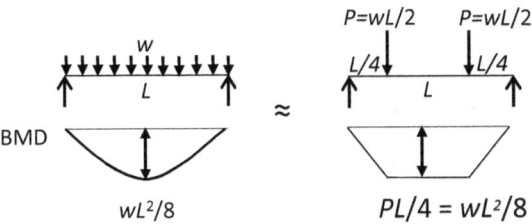

Thus PSC beams are usually type *I* or type *II*. So PSC beams typically have fewer links than RC beams not because prestressing force directly reduces the demand for links (as assumed by codes), but simply because PSC beams are typically <u>more slender</u> than RC beams.

4.3   DESIGN OF SHEAR REINFORCEMENT

In practice, we will use the <u>code</u> approach (e.g. EC2). Although usually perfectly safe, recent research has shown that for members with a point of contraflexure (PC) we should carry out an additional check in this region. For this use the "<u>Compressive Force Path</u>" approach (Kotsovos 2014). In both approaches shear is checked only at <u>Ultimate Limit State</u>. Thus we are only concerned with <u>safety</u>.

In order to represent line *ab* (see figure) EC2 uses the expression:

$$v_{fail} = v_{Rd,c} = 0.12k(100\rho_l f_{ck})^{1/3} + 0.15\sigma_{cp}$$

where

v_{fail} = nominal shear stress = $V_{fail}/b_w d$, and $V_{fail} a_v = M_{fail}$

k = size effect factor = $1 + (200/d)^{0.5} \le 2.0$ (where d is in mm)

ρ_l = longitudinal reinforcement ratio

       = $A_{sl}/b_w d$ ≤ 0.02

A_{sl} = area of tensile reinforcement extending at least $l_{bd}+d$ beyond section considered.

l_{bd} = design anchorage length

b_w = smallest width of cross section in tensile area.

EC2 distinguishes between two different types of shear failure:

1. Collapse initiated by <u>web-shear</u> crack (inclined crack which forms independently of any flexural cracks, and usually in a region of low moment).

2. Collapse initiated by <u>flexure-shear</u> crack (inclined crack which propagates from tip of flexural crack-similar to RC).

Thus there are two formulas for $v_{Rd,c}$ (a.k.a. v_{fail} i.e. shear strength of member without links). The two formulae are not related so we must check the shear using both of these formulas. If the tendon is profiled then we must check the shear at intervals along the span.

4.3.1   Web-Shear Crack

Occurs mostly in members having thin webs, such as I-sections. Normally it forms near the support where the beam is uncracked in flexure. Initially at least, the crack does not extend to the soffit of the beam. A typical crack of this type is shown in Fig. 4.5.

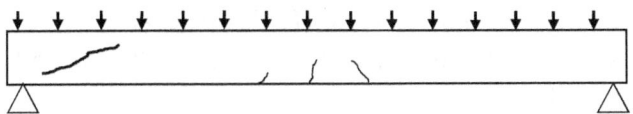

Fig. 4.5: Web-Shear crack

$V_{Rd,c} = (b_w I / A\hat{y})(f_{ctd}^2 + \alpha_1 \sigma_{cp} f_{ctd})^{0.5}$

where,

b_w = width of web (mm)

I = second moment of area of beam (mm^4)

Aŷ = first moment of area of the part of the section above the centroid (mm^3)

f_{ctd} = tensile strength of concrete (N/mm^2)

σ_{cp} = compressive stress <u>at centroid</u> due to prestress,

i.e. = $\gamma_p P_e/A$ (N/mm^2 where γ_p is 0.9)

α_1 = 1 for post-tensioned beams (generally less for pre-tensioned beams).

It is possible for web-shear crack to form in RC beams too. However it is much more likely to form in PSC beams because of the thinner webs used.

4.3.2   Flexure-Shear Crack

This crack usually occurs in the region of higher moment. The crack propagates from tip of a flexural crack. It is thus similar to that commonly seen in RC beams. It is the left-most crack in Fig. 4.6.

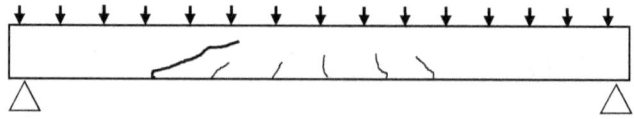

Fig. 4.6: Flexure-Shear crack

The shear strength of concrete section in this cracked region is (as noted previously):

$$V_{Rd,c} = [0.12k(100\rho_l f_{ck})^{1/3} + 0.15\sigma_{cp}]b_w d$$

where the terms are as noted previously.

The lower limit for $V_{Rd,c}$ is

$$[0.035k^{3/2}f_{ck}^{1/2} + 0.15\sigma_{cp}]b_w d$$

The upper limit for shear stress is $V_{Rd,max(22)}$ = $\alpha_{cw}0.124b_w d(1-f_{ck}/250)f_{ck}$

where $\alpha_{cw} = 1 + 1.5\sigma_{cp}/f_{ck}$

It is good practice to always provide at least nominal links.

4.4 COMPRESSIVE FORCE PATH APPROACH

This concept was originally proposed in the 1960s (Bobrowski and Bardhan-Roy 1969) but it has only recently been fully developed and experimental evidence obtained (Kotsovos 2014).

It is based on a simple idea: that the crack pattern just before failure reveals the path taken by compressions in the concrete and it is the strength of the concrete along this path that is crucial. Since concrete is weak in tension, it is tensile stresses developing along this "Compressive

Force Path" (the path of the compressions) that are the cause of shear failure.

Consider again a simply supported beam with equal symmetrical loads. See Fig. 4.7.

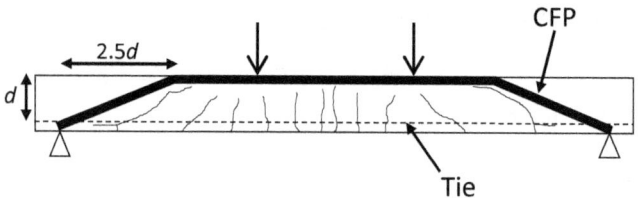

Fig. 4.7: Simply supported beam

The Compressive Force Path is revealed by the uncracked portion. Since this uncracked portion retains its stiffness tied-arch behaviour is suggested. When we consider a continuous beam, which consists of simply supported beams joined together at the points of contraflexure, it can be seen that this area is vulnerable as its strength depends on the tensile capacity (see Fig.4.8). The two compression booms require a tensile force to maintain equilibrium.

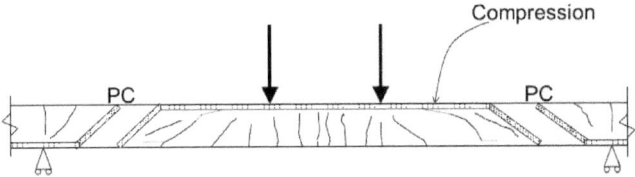

Fig. 4.8: Continuous Beam

4.4.1 Checking Capacity of PC region

Calculate ultimate shear force V_{ULT} at the PC from the ultimate load (i.e., load which would cause <u>flexural</u> failure); tensile resistance amounting to this force must be provided; i.e. provide links to resist this force,

i.e., $A_S = V_{ULT}/f_{yd}$;

Place these links within a length of $2h$ centered on the PC. This check should be carried out in addition to the code checks.

4.4.1.1 Example 1: Code Approach

A post-tensioned, simply supported rectangular beam spans 10 m (see below). The tendon profile is parabolic with midspan eccentricity 250 mm. Ultimate (factored) U.D.L. = 50 kN/m; $f_{ck} = 40$ N/mm^2 ($f_{ctk} = 2.5$ N/mm^2).

Using EC2 design shear reinforcement for section 1 m from the support.

Section properties:

$I = 5.4 \times 10^9$ mm^4 ;

$A = 180 \times 10^3$ mm^2;

$P_e = 700$ kN;

$A_p = 750$ mm^2

300

600

$e_m = 250$ mm

$e_e = 0$

10 m

Solution:

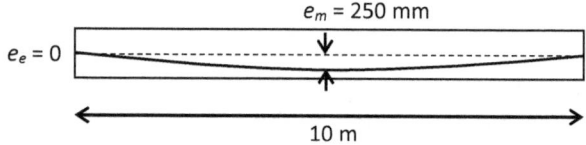

50 kN/m

1 m

10 m

250

250

200

SFD (kN)

Equation of parabola is: $y = ax^2 + bx + c$

For given tendon profile, following "boundary conditions" available:

at $x = 0$; $y = 0$ $(=> c = 0)$

at $x = 5$ m; $y = 0.25$ m $(=> 0.25 = 25a + 5b)$

at $x = 10$ m; $y = 0$ $(=> 0 = 100a + 10b)$

=> $y = -0.01x^2 + 0.1x$, is the equation of our parabola

Thus at $x = 1$ m, tendon eccentricity is $e = y = 0.09$ m

So, $d = 300 + 90 = 390$ mm

Flexure-shear:

$V_{Rd,c} = [0.12k(100\rho_l f_{ck})^{1/3} + 0.15\sigma_{cp}]b_w d$

where

$d = 390$ mm, $b_w = 300$ mm

$k = 1+(200/d)^{0.5} = 1+(200/390)^{0.5} = 1.72$ (≤ 2.0)

$\rho_l = A_p/b_w d = 750/(300 \times 390) = 0.006$ (≤ 0.02)

$\sigma_{cp} = \gamma_p P_e/A = 0.9 \times 700/180 = 3.5$ N/mm^2

$V_{Rd,c} = [0.12 \times 1.72(100 \times 0.006 \times 40)^{1/3} + 0.15 \times 3.5]300 \times 390$

$= [0.60+0.53]117\,000 = 132$ kN

The lower limit for $V_{Rd,c}$ is

$[0.035k^{3/2}f_{ck}^{1/2} + 0.15\sigma_{cp}]b_w d$

$= [0.035 \times 1.72^{3/2} \times 40^{1/2} + 0.15 \times 3.5]300 \times 390$

$= [0.50+0.53]300 \times 390 = 120$ kN < 132 kN

Web-shear:

$V_{Rd,c} = (b_w I/A\hat{y})(f_{ctd}^2 + \alpha_1 \sigma_{cp} f_{ctd})^{0.5}$

where,

I $\quad$ = 5.4×10^9 mm^4

$A\hat{y}$ $\quad$ = $300 \times 300 \times 150 = 13.5 \times 10^6$ mm^3

f_{ctd} $\quad$ = $f_{ctk}/1.5 = 2.5/1.5 = 1.67$ N/mm^2

σ_{cp} $\quad$ = 3.5 N/mm^2

α_1 $\quad$ = 1.0 for post-tensioned beams

$V_{Rd,c} = (300 \times 5.4 \times 10^9/13.5 \times 10^6)(1.67^2 + 1.0 \times 3.5 \times 1.67)^{0.5}$

$= 120000\sqrt{(2.79+5.85)} = 352$ kN > 132 kN

$=> V_{Rd,c} = 132$ kN

Shear strength of this section is:

$V_{Rd,c} + \gamma_p P_e \, sin\beta$

where $V_{Rd,c}$ = shear strength of concrete; P_e = effective prestress force; β = inclination of tendon to horizontal. ($P_e sin\beta$ is vertical component of prestress).

Now $sin\beta$ is roughly equal to $tan\beta$ (since angles small)

$y = -0.01x^2 + 0.1x$, is equation of parabola

$=> dy/dx = -0.02x + 0.1 = 0.08$ at x = 1 m

$=> P_e sin\beta = 700(0.08) = 56$ kN at x = 1 m

Thus the shear strength = $V_{co} + \gamma_p P_e \, sin\beta = 132 + 0.9 \times 56$
= 182 kN

The upper limit for shear stress is:

$V_{Rd,max(22)} = \alpha_{cw}0.124b_wd(1-f_{ck}/250)f_{ck}$

where $\alpha_{cw} = 1 + 1.5\sigma_{cp}/f_{ck} = 1 + 1.5\times3.5/40 = 1.13$

$V_{Rd,max(22)} = 1.13\times0.124\times300\times390\times(1-40/250)40$

$= 551$ kN > 250 kN

- Design links: Try H8 links (2 legs):

$A_{sw}/s = V_{Ed}/(0.9d0.87f_{yk}\cot\theta)$

$100.5/s = (200-0.9\times56)\times10^3/(0.9\times390\times0.87\times500\times2.5)$

$\Rightarrow$ $s = 100.5/0.39 = 257$ mm

$A_{sw}/s = 100.5/257 = 0.39$

- Nominal links: Try H8 links (2 legs):

$A_{sw,min}/s = 0.08f_{ck}^{1/2}b_w/f_{yk}$

$= 0.08\times40^{1/2}\times300/500 = 0.30 < 0.39$

Provide H8 links at 250 mm

This is the code required reinforcement at and around x = 1m.

In practice further checks on shear would normally be performed every 1m or so along span.

Check using "Compressive Force Path":

As this member is a single span, simply supported member, it has no point of contraflexure. Thus there is no additional reinforcement required according to Compressive Force Path. Thus code approach is sufficient here.

4.4.1.2   Example 2

The following post-tensioned roof beam is 300mm wide and 600mm deep. It supports a uniformly distributed *DL* = 5 kN/m in addition to its self-weight, and a *LL* = 6.6 kN/m (both loads over its full length).

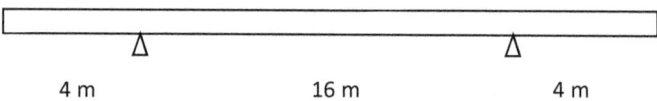

4 m   16 m   4 m

Post-tensioning is provided to meet the flexural serviceability, transfer, and ultimate requirements of EC2. The beam area and material strengths are as example 1. Shear reinforcement is provided for beam according to the requirements of EC2. What additional reinforcement is required at points of contraflexure according to Compressive Force Path?

Solution:

Beam s/wt = 0.3x0.6x25 = 4.5 kN/m. Ultimate *UDL* to be supported by beam is 1.35(4.5+5)+1.5(6.6) = 22.7 kN/m

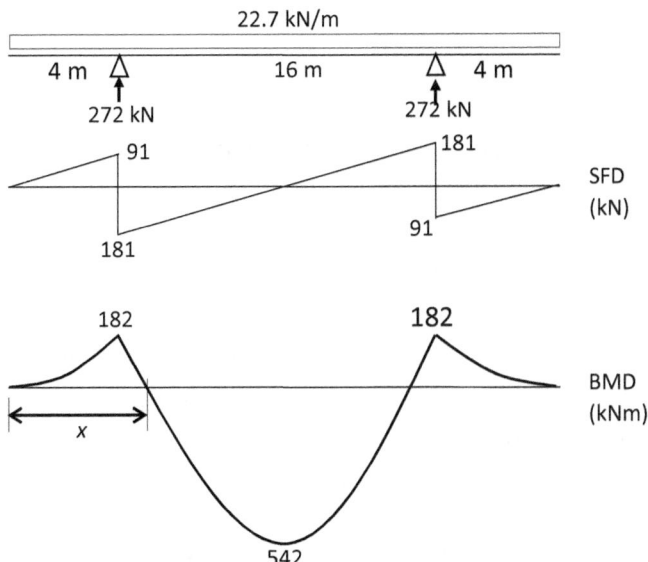

At Point of Contraflexure *BM* =0:

Thus $22.7x^2/2 - 272(x - 4) = 0$, Hence $x = 5.07$ m

Shear at Point of Contraflexure:

= 22.7(5.07)-272 = 157 kN

Provide links to resist ultimate force of 157 kN. Spread these links over 2*h*.

Area of reinforcement required = 157,000/(0.87x500) = 361 mm^2.

Number of 2 leg H8 links (A_{sw} = 100.5 mm^2) = 3.6 sets.

Code nominal links H8@250 from example 1 provides 2x600/250 = 4.8 sets.

Thus code nominal links are sufficient; no extra links are required.

4.5 CURTAILMENT AND ANCHORAGE

4.5.1 Terminating Tendons

In pre-tensioning, it is common practice to deliberately "de-bond" some of the tendons near a (simple) support. This may be done by covering strands with a plastic pipe to prevent bond with the subsequently poured concrete. It is done to reduce the flexural stresses at the end of the beam as the moment due to deadweight is zero at the ends.

However, tests show that a <u>stress concentration</u> may develop at the point where the de-bonding begins. If too many tendons are de-bonded this can result in early cracking at that point and can lead to a *reduction* in shear strength. To avoid reducing the shear strength it is usual to avoid de-bonding more than 25% of the strands (as recommended by Abeles (1981)).

Where tendons are profiled, bending up strands near the end of a beam appears to cause no such stress concentration, so is a better way of controlling stresses near the ends of the beam. 100% of the strands may be bent-up near the ends, as such bending up does not introduce a stress concentration.

4.5.2   Short Tendons

In post-tensioning it is common to use "short tendons" if the end-span controls the design. The arrangement is shown in Fig. 4.9. Here the end-span is of equal length to the interior spans. This would result in larger moments in the end-span. The drape here would be lower than in interior spans, so it is likely the end-span controls the design. This tendon has a dead-end at roughly the quarter point of the interior span and is stressed from the left-hand end.

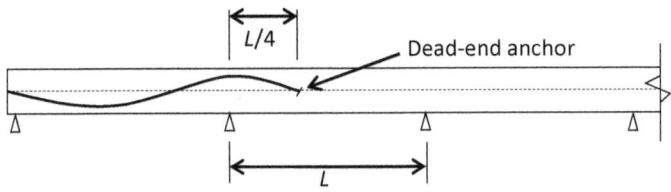

Fig. 4.9: Short tendon in end span.

If the extra tendon is terminated too close to the internal support then it acts as a stress concentration and can reduce the shear strength of the support region. The tendon should continue at least $L/4$ past the penultimate support. Thus it is usually past the point of contraflexure (typically at about $L/5$). Fortunately at $L/4$ the tendon is about mid-depth of the member so the anchor can readily be accommodated within the depth of the member.

CHAPTER 5: CONTINUOUS MEMBERS IN FLEXURE

5.1 ADVANTAGES OF CONTINUOUS BEAMS

- Design moments are smaller for given spans and loads;

- Deflections are reduced so a shallower section can be used;

- Use of continuous tendons saves end anchorages and stressing operations, so is more economical;

5.2 DISADVANTAGES

- Friction losses are larger when many reversed curves are used for the tendon profiles;
- Shortening of long continuous beams may produce excessive restraint forces in supports;
- Additional computational effort required especially due to complication of secondary moments caused by prestressing.

5.3   ANALYSIS

5.3.1 Serviceability load range:

Model the effect of prestressing force using **"equivalent loads"** (to be discussed later);

Conventional linear elastic analysis for indeterminate structures, i.e. as *RC*;

5.3.2   Ultimate load range:

Conventional linear elastic analysis method with limited moment redistribution, i.e., as *RC*.

5.4   EFFECT OF PRESTRESSING FORCE

5.4.1   Statically determinate structures

In statically <u>determinate</u> structures, e.g., simple beams, application of prestress force does *not* cause any reaction at supports. And so support reactions in statically determinate structures are determined *only* by the equilibrium of <u>externally</u> applied loads.

5.4.1.1. Illustration: simply supported beam

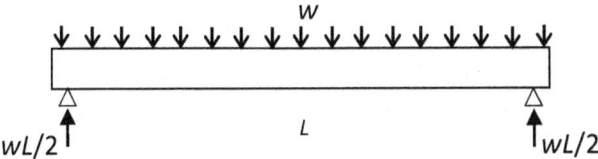

Consider a simply supported beam span L under a UDL w. The support reaction is $wL/2$ (neglecting the beam's self-weight). It depends *only* on span of beam and the UDL: it makes *no* difference to the reaction whether beam is prestressed or not, the degree of prestressing or what tendon profile is used.

5.4.2   Statically indeterminate structures

In statically <u>indeterminate</u> structures, e.g., continuous beams, prestress force will normally produce support reactions even when no gravity load is applied. Therefore, to calculate stresses, it is first necessary to find the reactions (and thus moments) due to the prestress force itself. This is the main complication when it comes to statically indeterminate structures.

5.4.3   Equivalent Load Approach

The effect of a prestressing tendon acting on the concrete is imagined as being equivalent to loads being put on the concrete. The beam is analyzed under effect of these loads using conventional beam/frame linear elastic analysis.

The shape that a cable would take under a given loading is the same as the shape of the bending moment diagram of a simply supported beam under that loading. This is nicely illustrated by a suspension bridge. The main load is the deadweight of the cables, hangers and deck. Thus a UDL is applied to the main cables. In response they take the shape of a parabola (actually a catenary under the cable weight but a parabola to a close approximation). Thus a UDL causes a cable to take the shape of a parabola. Similarly, under a point load the cable takes up the shape of two straight lines with the "kink" in the cable under the point load.

5.4.3.1   Illustration for UDL:

Beam:

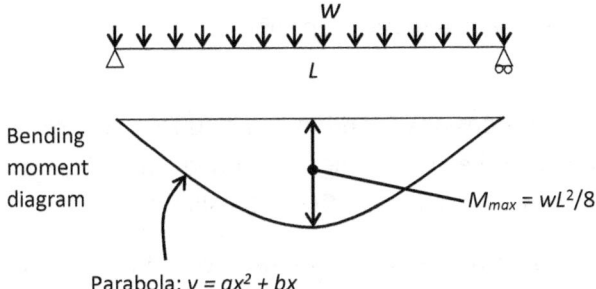

Bending moment diagram

$M_{max} = wL^2/8$

Parabola: $y = ax^2 + bx$

Now consider a chain under the same load:

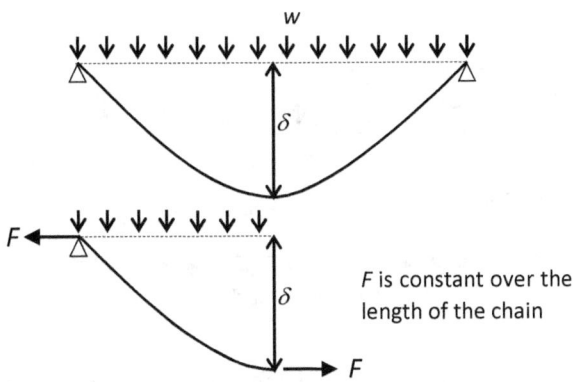

F is constant over the length of the chain

Statics requires the same moments to be resisted. Chains can resist tension only (no BM resistance). Thus the chain must be in the shape of a parabola.

$M_{max} = wL^2/8 = F\delta$, where F is chain tension and δ is the displacement of the tendon.

5.4.4   Relevance to post-tensioning?

Thus a post-tensioned tendon, which is <u>forced</u> to remain parabolic by the surrounding hardened concrete, exerts a *UDL* on the concrete when it is tensioned.

Similarly a harped tendon (a tendon consisting of straight lines with a "kink" where they intersect) exerts a point load on the concrete at the kink point of the tendon.

5.4.4.1   *Parabolic tendon–simple/continuous beam*

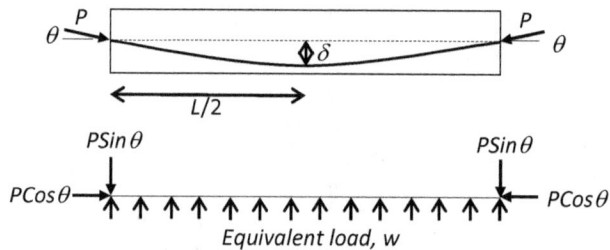

From above, $wL^2/8 = F\delta$, where F is chain tension and δ is the mid span displacement of the chain,

Thus: $wL^2/8 = P\delta$ => Equivalent upward UDL = **$w = 8\,P\delta/L^2$**

Note: The upward load changes as P changes.

- Initial value of prestress P_i, is used for the check on stresses at transfer => w_i

- Effective prestress P_e is used for the service check => w_e

Dimension δ (known as "drape" of tendon) is **vertical** distance measured from ends of tendon to its position at midspan. It is *not* necessarily measured relative to the beam centroid.

5.4.4.2   *Parabolic tendon-cantilever*

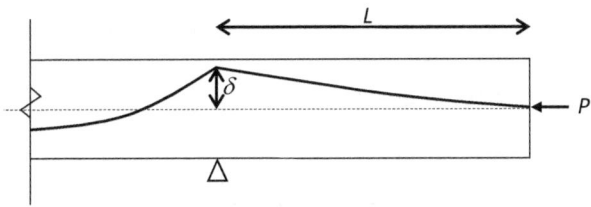

Equivalent load, w

For full simply supported span considered above we know $P\delta = "wL^2/8"$, i.e. the *static* moment.

Moment at root of cantilever = $wL^2/2$

Thus, by analogy, $P\delta = wL^2/2$, i.e. $\boldsymbol{w = 2P\delta/L^2}$

5.4.4.3 Harped tendon–simple/continuous beam

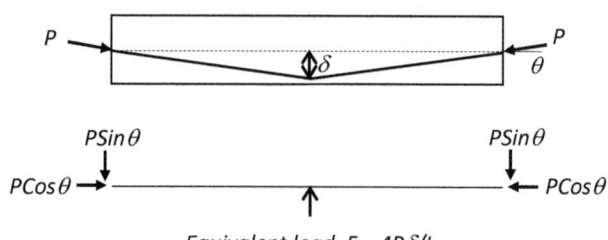

Equivalent load, F = 4PδL

$$W = 2Psin(\theta) \approx 2P(\theta) = 2P(\delta)/(L/2)$$

A point load W placed at the mid-span of a simply supported beam causes a $M_{max} = WL/4 => P\delta = WL/4 => $ **$W = 4P\delta/L$**

5.4.4.4 Harped tendon-cantilever

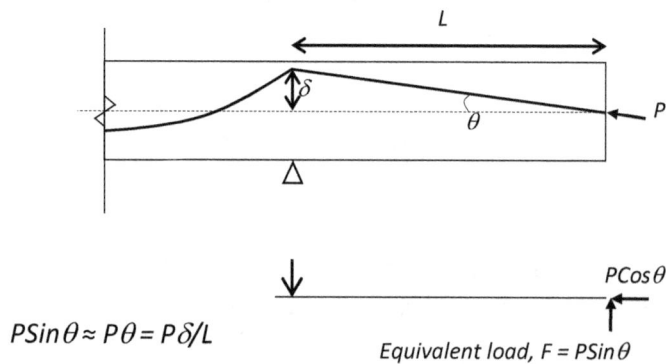

$$PSin\theta \approx P\theta = P\delta/L$$

Equivalent load, F = PSin θ

5.4.4.5   Straight tendon

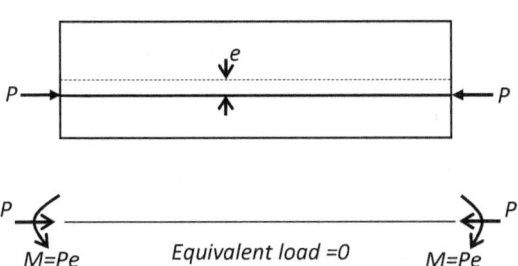

M=Pe   Equivalent load =0   M=Pe

5.4.4.6   Examples of equivalent loads in various cases

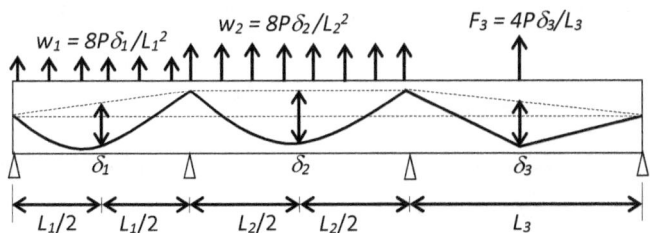

In the case of spans 1 and 3 the load is actually at a small angle to the vertical, but as the span/depth for real members is usually high, we can ignore this angle.

5.4.5   Total, Primary and Secondary moments

As mentioned above, equivalent load concept allows us to visualize effect of prestressing in terms of loads applied to beam. To find effect of the prestressing we must analyze the beam *under these loads*. This analysis gives all the bending moments caused by the prestress. We will call these *total* moments (M_{TOT}). We divide this total moment into *primary* and *secondary* contributions.

- Primary moment *(M_1):*

That part of total moment equal to magnitude of prestressing force *P*, multiplied by the eccentricity from centroid *e* at that section.

- Secondary moment *(M_2):*

Remainder of total moment.

It is the moment induced by the *reactions* that result from prestressing.

- $M_{TOT} = M_1 + M_2$

Why is it necessary to distinguish between primary and the secondary effects?

Because of the presence of secondary moments prestressing alters the reactions of the beam. This effect is present up to ultimate collapse. Thus we need to take these secondary moments into account when we calculate the ultimate demand on a section.

5.4.5.1 Illustration: secondary moment

Consider a weightless beam Fig. 5.1(a). On prestressing typically the beam tries to rise off the interior support (b). However, if it is monolithically connected to the support it cannot freely lift off. Thus reactions are induced in order to hold the beam down (c).

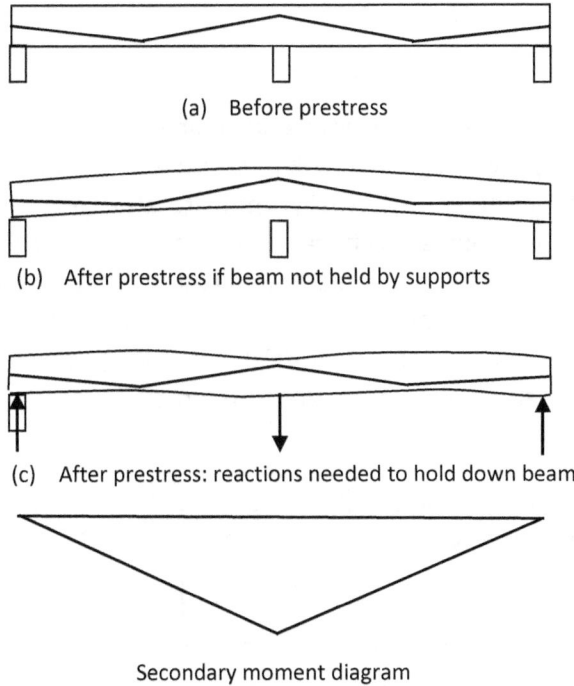

(a) Before prestress

(b) After prestress if beam not held by supports

(c) After prestress: reactions needed to hold down beam

Secondary moment diagram

Fig. 5.1: Secondary moments in two span beam

(If the tendon profile is what's termed 'concordant' there is no M_2 but this profile is inefficient at ultimate and so is rarely used; a typical tendon profile will result in the beam trying to lift-off the middle support.)

Although called "secondary" these moments are often not small. Secondary moments are usually sagging: thus the hogging moment at supports is <u>reduced</u>. However, the sagging moment in the span is <u>increased</u>.

5.4.6   End Column

Moments are reduced in the end-column of a frame (see Fig. 5.2). Although the columns are not prestressed directly (they are RC usually) they are attached to the beams or columns which are prestressed. Therefore they rotate with these elements. Thus all the column moment is secondary moment. The prestressing causes an upward load on the beam/slab, so the gravity load induced column moments are counteracted.

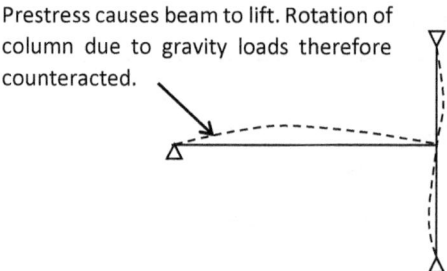

Prestress causes beam to lift. Rotation of column due to gravity loads therefore counteracted.

Fig. 5.2: End-column in a prestressed frame

5.4.7   Secondary Moments in Continuous Beams

Consider a cut through a weightless beam. The simple support of a statically indeterminate beam is shown in Fig. 5.3 below.

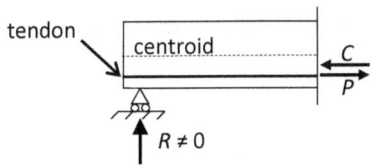

Fig. 5.3: End support of continuous beam

$P = C$ for horizontal equilibrium, but now consider rotational equilibrium. As $R{\neq}0$ it means C and P cannot be on the same line. (Compare Fig. 2.2).

5.4.8   Summary

Analysis:

- Total moment due to prestressing in continuous beams is the sum of primary and secondary moments i.e., $M_{TOT} = M_1 + M_2$.

- Using "equivalent load method" the total moment M_{TOT} is found first.

- Primary moment is found as $M_1 = P.e$

- Secondary moment $M_2 = M_{TOT} - M_1$.

Obtaining stresses:

- Compute total moment due to prestressing M_{TOT} using the equivalent load method;
- Compute stresses due to M_{TOT}.
- Add stresses due to applied gravity loads.

Note that the stresses due to prestressing depend only on M_{TOT} .

5.4.8.1   Example 1

A two-span continuous rectangular beam, has a total depth h of 600 mm and a width b of 300 mm. The tendon is parabolic and midspan and support eccentricities are as shown below. The prestressing force after losses, P_e = 900 kN.

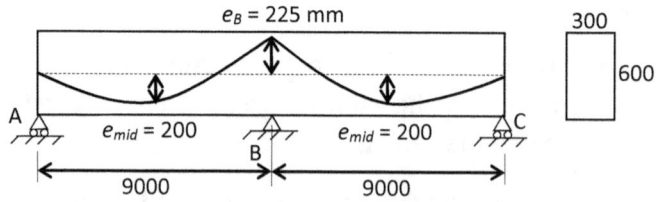

Using the idealized tendon profile shown, calculate the following:

(a)   primary and secondary moments at B,

(b)   top fibre concrete stresses at support B due to prestressing only,

(c)   the top fibre concrete stresses at support B due to prestressing plus a *UDL* of 15 kN/m (including self-weight) applied on both spans.

Section data: A_c= 180,000 mm^2; I_c = 5.4 $(10)^9$ mm^4; Z_c = 18 $(10)^6$ mm^3

Solution

Consider span AB: the equivalent upward load:

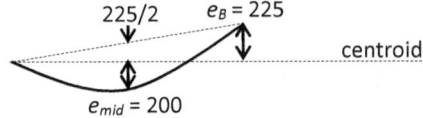

225/2 e_B = 225
e_{mid} = 200
centroid

For span AB, the parabolic tendon has a total drape δ of 200 + 112.5 = 312.5 mm. So the equivalent load is a *UDL* equal to $w = 8P_e\delta/L^2$.

$w = 8(900)(0.3125)/(9)^2 = 27.7$ kN/m

Thus, the total effect of prestressing is *exactly* the same as the effect of applying a *UDL* of 27.7 kN/m upward (plus the horizontal force P_e).

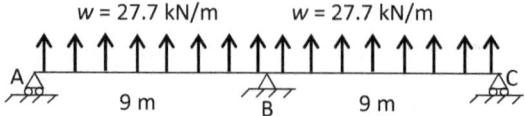

The moment from this equivalent load is the <u>Total</u> moment due to prestressing (M_{TOT}).

And $M_{TOT} = M_1 + M_2$

Due to symmetry of geometry and loading,

M_{TOTB} $= wL^2/8 = (27.7)(9)^2/8 = 280$ kNm (Sagging)

    (a) The <u>primary</u> moment at B

$M_1 = P.e = 900 \times 0.225 = 202$ kNm   (Sagging)

Thus the <u>secondary</u> moment at B

$M_2 = M_{TOT} - M_1 = 280-202 = 78$ kNm   (Sagging)

    (b) Top fibre stress at B (due to prestressing)

$f_2 = P/A_c + M_T/Z = 900/180 + 280/18 = 5 + 15.6$

$= 20.6$ N/mm^2 (i.e., compression)

    (c) Top fibre stress at B (due to prestressing and gravity UDL of 15 kN/m)

From symmetry of UDL and beam

$M_B = wL^2/8 = 15(9)^2/8 = 152$ kNm,

$f_2 = 20.6 - 152/18 = 20.6 - 8.4$

= 12.2 N/mm² (i.e., compression)

5.4.8.2   Example 2

Repeat the previous example with beam spans 9 m and 12 m. The idealized tendon profile is as shown below.

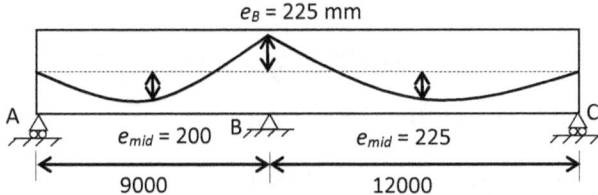

As before, the prestressing force after losses, P_e = 900 kN

(a) Calculate primary and secondary moments at B,

(b) The top fibre concrete stresses at support B due to prestressing only,

(c) The top fibre concrete stresses at support B due to prestressing plus a UDL of 15 kN/m (includes self-weight) applied throughout both spans.

Section data: A_c= 180,000 mm² ; I_c= 5.4 (10)⁹ mm⁴ ; Z_c = 18 (10)⁶ mm³

Solution

Equivalent upward load:

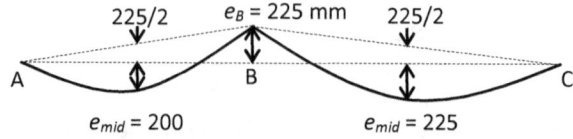

$225/2$ $e_B = 225$ mm $225/2$

A   B   C

$e_{mid} = 200$ $e_{mid} = 225$

For AB, $\delta = 200 + 112.5 = 312.5$ mm.

For BC, $\delta = 225 + 112.5 = 337.5$ mm

Both correspond to an equivalent load $w = 8P_e\delta/L^2$.

$w_{AB} = 8(900)(0.3125)/(9)^2 = 27.7$ kN/m

$w_{BC} = 8(900)(0.3375)/(12)^2 = 16.9$ kN/m

Thus, the total effect of prestressing is the same as the effect of applying the following loads.

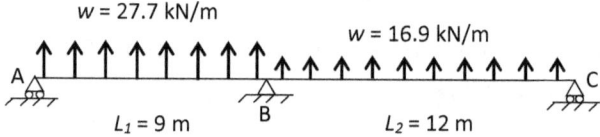

$w = 27.7$ kN/m

$w = 16.9$ kN/m

A   B   C

$L_1 = 9$ m $L_2 = 12$ m

The beam is analyzed using any appropriate technique. The analysis using moment distribution as follows.

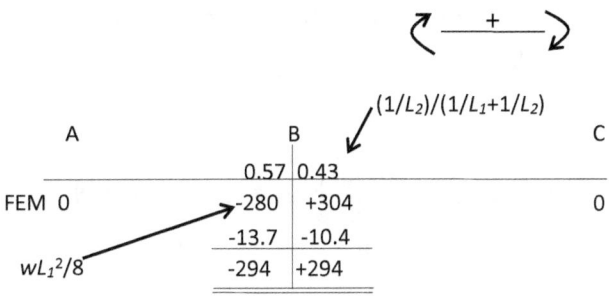

The moment from this equivalent load is the <u>Total</u> moment due to prestressing (M_{TOT}). And $M_{TOT} = M_1 + M_2$

From moment distribution,

M_{TOTB} = 294 kNm (Sagging)

(a) The Primary moment at B

$M_1 = P.e = 900$x$0.225 = 202$ kNm (Sagging)

Thus the Secondary moment at B

$M_2 = M_{TOT} - M_1 = 294 - 202 = 92$ kNm   (Sagging)

(b) Top fibre at B (due to prestressing only.)

$f_2 = P/A_c + M_{TOT}/Z = 900/180 + 294/18$

$= 5 + 16.3 = 21.3$ N/mm^2 (i.e., compression)

(c) Top fibre at B (due to prestressing and UDL = 15 kN/m).

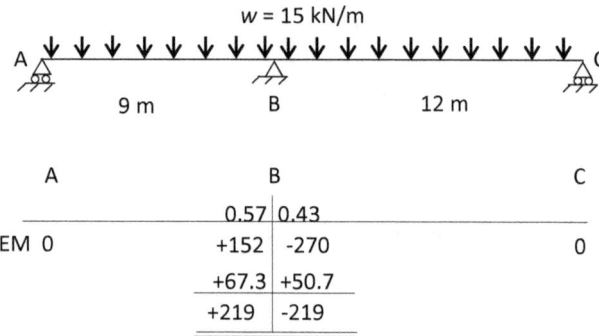

A	B		C
	0.57	0.43	
FEM 0	+152	-270	0
	+67.3	+50.7	
	+219	-219	

Use superposition:

$f_2 = 21.3 + M_B/Z = 21.3 - 219/18 = +9.13$ N/mm^2
(i.e. compression)

5.5 DESIGN USING LOAD BALANCING

Load balancing is a very simple but very useful concept. Based on equivalent loads, the method can be used for the preliminary design for members with tendon profiles that are <u>not</u> straight, e.g., parabolic, harped. Not surprisingly, it is used mainly for design of post-tensioned members. It makes design of continuous beams and slabs easier.

As we have seen, a parabolic tendon exerts a *UDL*. Thus we can counteract effect of a **downward** *UDL* using a parabolic tendon. For example, if the downward load due to effects of *DL* and *LL* is say 15 kN/m and the upward load due to the tendons is 15 kN/m then the whole load is balanced by the tendons.

When this 15 kN/m downward load is applied the beam experiences *no bending* at all, only compression due the axial effect of the tendons, i.e., the stress everywhere is P_e/A. The beam would have *no deflection* under this load. The prestressing is said to have "balanced" the gravity loading. (Of course load hasn't really disappeared (reactions do not change unless the beam is statically indeterminate) but as far as the beam is concerned, *any bending stresses have gone*.) If a downward load of 20 kN/m is applied then only **unbalanced** load, i.e., 20-15 = 5 kN/m, causes bending stresses in beam.

The design decision then really becomes one of *what load should be balanced?*

It is common to balance all sustained (i.e., quasi-permanent) load on an element. Thus entire *DL* often balanced, as is any additional *DL*, e.g. screed. A portion of *LL* is sometimes balanced. The amount of *LL* included depends on type of structure, e.g. Residential 0-20%; Offices 0-30%; Storage facilities 0-60%.

5.5.1 Where has the balanced load gone?

Nowhere! These are <u>internal</u> loads. Consider parabolic tendons: The equivalent *upward* UDL is $w = 8P_e\delta/L^2$ This is the load exerted <u>by the steel on the concrete</u>. An equal and opposite *downward* UDL is exerted <u>by the concrete on the steel</u>. The result is to impose a *deformation*, which when the structure is determinate, causes no reactions. Thus there is no *net* load (i.e. external) on the beam.

The above illustration used parabolic tendons as an example. However the principle applies to any non-straight (over full span of member) tendon profile, i.e., any harped tendon profile.

5.5.1.1   Example: Design of Simply Supported Beam

A 300 mm wide 1000 mm deep post-tensioned beam spans 16 m. $DL = 15$ kN/m (including self-wt.) and a $LL = 10$ kN/m. Use load balancing to choose a suitable value of effective prestress P_e required if tensile stresses limited to 4 N/mm².

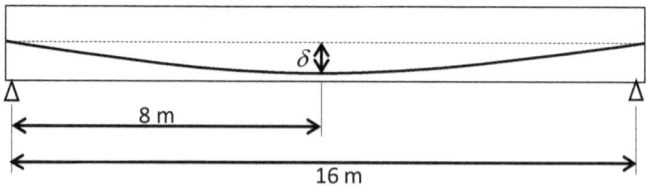

Solution

Let us choose to **balance all** of DL, i.e., 15 kN/m.

We can choose a suitable value for δ by observing cover requirements. Suppose we keep the cover to the centre of the tendon at 100 mm at midspan then δ is 400 mm.

Then writing our formula in terms of P_e we get

$P_e = wL^2/8\delta$

Thus effective value of prestress needed is $15\times16^2/(8\times0.4) = 1,200$ kN

i.e., when 15 kN/m is applied only stresses are axial compressive stresses, i.e., $P_e/A = 1200\times10^3/300\times1000 = 4$ N/mm^2. The bending stresses are due to the **unbalanced load** of 10 kN/m.

These are a maximum at midspan where the moment M is $10\times16^2/8 = 320$ kNm.

The tensile stress due to this is $f_1 = M/Z = 320\times10^6\times6/300\times1000^2 = 6.4$ N/mm^2.

Thus the net stress in the beam is -6.4 + 4.0

$= -2.4$ N/mm^2, i.e., tensile

But allowed a tensile stress of 4 N/mm^2 => okay.

Comment: Thus it can be seen that in this example the amount of prestress could be reduced (so that balanced load < DL) and stresses would still be okay. However, deflections are likely to increase. This is the main problem with choosing to balance less than *DL*.

5.5.1.1. Example: Design of Continuous Beam

A post-tensioned concrete rectangular beam consists of two 10 m spans. It is 300 mm wide and 600 mm deep. *DL* = 25 kN/m (including self-weight) and *LL* = 15 kN/m. Cover requirements mean that centre of the tendon should be no closer than 100 mm from the surface of the concrete.

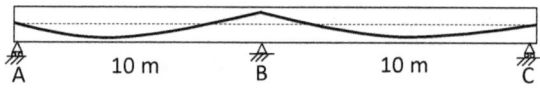

Calculate the following:

1.   Minimum prestress force required to balance all of dead load.

2.   Moment at B at service.

3.   Stresses at top and bottom of section at B at service.

4.   Stresses under **frequently** applied load (*DL*+ 50%*LL*)?

5.   Moment at B that is to be used in design for ultimate load conditions.

Solution

1.   Tendon profile parabolic:

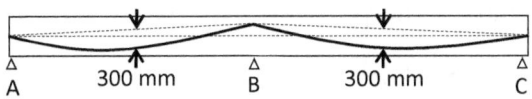

$\Rightarrow w_{bal} = 8P_e\delta/L^2$ where $\delta = 300$ mm

All of *DL* to be balanced by prestressing, i.e., require

$w_{bal} = 25$ kN/m.

This means we require $P_e = w_{bal}L^2/8\delta$

$= 25 \times 10^2/8 \times 0.3 = 1042$ kN

2.   Moment at B due to load:

Unbalanced load is 15 kN/m (downward).

$w = 15$ kN/m

A

$L = 10$ m

B

Thus unbalanced *BM* at B at service $= wL^2/8$

$= -188$ kNm   (i.e., hogging)

3.   Stresses at B

$\sigma_{TOP} = + P_e/A - M_{unbal}/Z$

$=+1042\text{x}10^3/(600\text{x}300)- 188\text{x}10^6\text{x}6/(300\text{x}600^2)$

$= +5.8 - 10.4 = -4.6 \text{ N/mm}^2$

$\sigma_{BOT} = + P_e/A + M_{unbal}/Z$

$= +5.8 +10.4 = +16.2 \text{ N/mm}^2$

4.   Frequent load:

$DL+ 50\%LL = 25 + (0.5)15 = 32.5 \text{ kN/m};$

Balanced load = 25 kN/m,

Thus **unbalanced** load = 32.5 – 25 = 7.5 kN/m

$M_{unbal} = wL^2/8 = 7.5\text{x}10^2/8 = 93.8 \text{ kNm}$ (hogging)

$\sigma_{TOP} = + P_e/A - M_{unbal}/Z = + 1042\text{x}10^3/(600\text{x}300) - 93.8\text{x}10^6\text{x}6/(300\text{x}600^2) = +5.8 - 5.2 = +0.6 \text{ N/mm}^2$

$\sigma_{BOT} = + P_e/A + M_{unbal}/Z = +5.8 +5.2 = +11.0 \text{ N/mm}^2$

5.   Design Moment at Ultimate at B:

= Moment due to $(1.35DL+1.5LL)$ + (worst of 1.1 or 0.9)M_2

=> must find secondary moment at B, i.e., M_{2B}

$M_{TOTB} = w_{bal}L^2/8 = 25\text{x}10^2/8 = 313 \text{ kNm}$ (sagging)

$M_{1B} = P_e.e = 1042\times0.2 = 208$ kNm (sagging)

$M_{2B} = M_{TOTB}-M_{1B} = 313-208 = 105$ kNm   (sagging)

Design moment at ultimate:

$= - (1.35\times25+1.5\times15)\times10^2/8 + 0.9\times105$

$= - 703 + 94 = - 609$ kNm (i.e., hogging)

Note: Compare this moment (609) with that if beam were conventionally reinforced and **not** prestressed, i.e., **703 kNm** (hogging).

5.5.2   Ultimate considerations

As stated previously, the primary moment due to prestress, M_1, does *not* change the reactions. It is an internal force not an external one, so there is no need to consider it at *ULS*. Similarly, the axial force due to prestressing *not* an external force (it is equilibrated at the anchorage) so it too should not be considered at the *ULS*.

The secondary moment, M_2, results in reactions which persist to *ULS* so it is correct to include M_2 at the *ULS*.

There is no increase in the reactions due to prestress at the *ULS* (only gravity loads are factored and M_2 is independent of them) so the load factor is around 1.0 (EC2 says take worst of 0.9 and 1.1 to be conservative).

5.5.2.1   *Illustration*

Consider a post-tensioned cantilever as perhaps it is the clearest illustration. The primary moment is a moment on the concrete. But as C = P there is an equal and opposite set of moments counteracting this primary effect. Thus there is no net moment to be considered at ultimate. This accords with our 'common sense'. Otherwise if the upward load due to the tendon equalled the downward load at ultimate due to gravity loads (i.e. 1.35DL+1.5LL), then the beam would be required to have no strength at all!

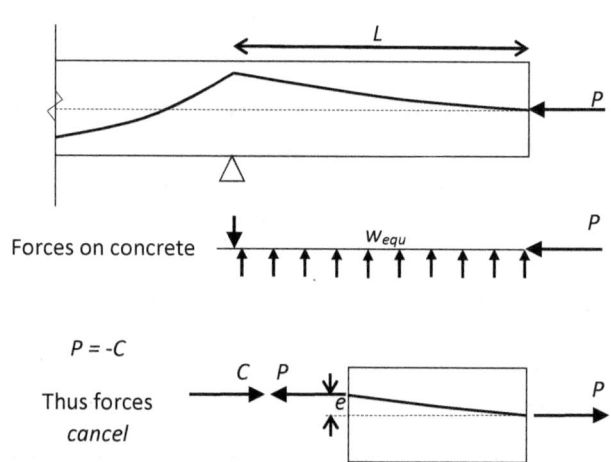

Forces on concrete

$P = -C$

Thus forces
cancel

5.5.2.2   Example: Design of simply supported Transfer Beam

A transfer beam is 500 mm wide and 1000 mm deep. It supports a column at midspan and a UDL (*DL*= 15 kN/m including self-weight, *LL* = 10 kN/m). Assume cover to the centreline of tendon is 100 mm from the surface.

(a)   Suggest a suitable tendon profile to balance the DL of the point load,

(b)   Estimate prestressing force required,

(c)   Calculate midspan stresses at the bottom of the beam due to the unbalanced load.

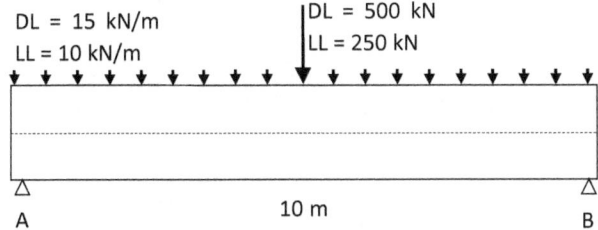

DL = 15 kN/m
LL = 10 kN/m

DL = 500 kN
LL = 250 kN

A   10 m   B

Solution

(a)   The tendon profile appropriate to balancing a point load is a one point depression tendon ("harped" tendon); the depression is at midspan. The tendon is otherwise straight and terminates at mid depth at the supports.

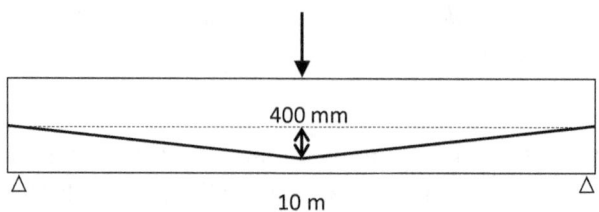

(b) Balance entire *DL* of point load, i.e. 500 kN.

Upward load induced by this harped tendon is $W = 4P_e\delta/L$ (see section on equivalent load).

Thus $P_e = WL/4\delta = 500 \times 10/4 \times 0.4 = 3125$ kN

(c) The unbalanced loadings are as follows:

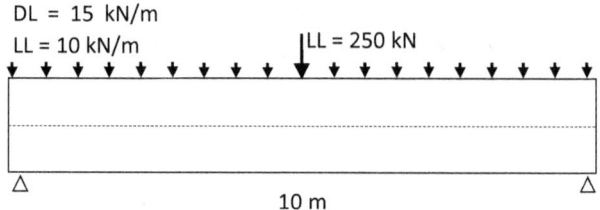

Unbalanced midspan moment due to a point load and a distributed load:

$M_{UNBAL} = (P_{UNBAL}L/4 + w_{UNBAL}L^2/8)$

$M_{UNBAL} = 250 \times 10/4 + (10+15) \times 10^2/8 = 938$ kNm

Midspan stresses at bottom:

$f_1 = M_{UNBAL}/Z + P_e/A = -(938 \times 10^6) \times 6/500 \times 1000^2$

$$+ 3125 \times 10^3/1000 \times 500$$

$= -11.3 + 6.25 = -5.05 \text{ N/mm}^2 \qquad \text{(tension)}$

5.5.2.3   Summary of Design Method (tendon straight):

(1)   Start with trial value of section size and trial tendon layout.

(2)   Check *ULS* Moment.

(3)   Estimate Losses => P_i and P_e known.

(4)   Check Service Stresses.

(5)   Check Transfer Stresses.

(6)   Check *ULS* Shear.

(7)   Check *SLS* Deflection & Vibration.

(8)   Detail member.

5.5.2.4   Summary of Design Method (tendon non-straight):

(1)   Use Load Balancing to choose section size and P_e required for the given *SLS* stresses. Normally balance "sustained" load.

(2)   Estimate Losses => P_i and P_e known.

(3)   Check Transfer Stresses.

(4)   Check *ULS* Moment.

(5)   Check *ULS* Shear.

(6)   Check *SLS* Deflection & Vibration.

(7)   Detail member.

The suggested order is for convenience only.

Note: Usually the <u>ideal</u> tendon profile is used for preliminary calculations, i.e. the effect of smoothing the tendon profile over supports is usually neglected at this stage. (This is usually conservative for the prestress force.) The 'real' profile is shown below.

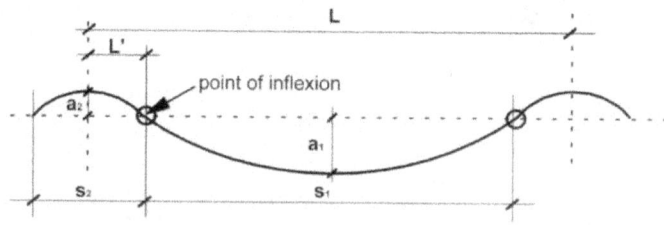

Typically, $L'/L = 0.1$ and $a_1/(a_1 + a_2) = 80\text{-}90\%$

Hence the rule of thumb: For upward equivalent load:

"Take 90% of the drape over 90% of the span"

CHAPTER 6: FLOOR SYSTEMS

6.1 COMMON SYSTEMS:

- One-way systems: beam and one-way slab.

    - may be *in situ* or precast;

    - downstand increases formwork turnaround time;

    - partition layouts fixed by presence of downstand;

    - services layouts complicated by presence of downstand.

- Two-way systems: flat slab/flat plate.

    - *in situ* only (i.e. post-tensioned only);

    - likely to be lowest formwork cost

    - holes near columns difficult;

    - deflection higher at edges

    - heavy solution.

Generally, if the panel's aspect ratio > 1.4 a one-way system is likely to be more economic. For more information about optimizing floor layouts refer to Zahn (1992).

6.1.1   Design

• One-way:

Design beam and slab using EC2 as 'cracked' elements (see section 2).

Typically P_e/A = 1-2 N/mm² (slab)

and 5 N/mm² (beam).

• Two-way:

Design using Concrete Society Technical Report TR43.

Typically P_e/A = 1-2 N/mm².

Flexural cracking anticipated so the sections are considered as 'cracked'.

6.1.2   Structural Analysis of Two-way system

The horizontal structure (slab/beams) is post-tensioned; the vertical structure is usually not. The analysis is as for conventional RC: i.e., for regular arrangements of columns it usually based on the Equivalent Frame Method (EFM).

Special software written for post-tensioning (e.g., RAPT, ADAPT) is useful but not essential. We may use a Finite Element (or grillage) program for very complex floors, or when repetition is extensive (e.g. SAFE). But we certainly should not use the EFM if the floor is very irregular (e.g., if columns > 10% of typical span off-line).

It is good practice to ensure the construction is <u>braced</u>. Otherwise (1) P-delta effects become large as the structure is relatively heavy and (2) the connection between the column and the slab highly stressed.

The following relates to the EFM. Member properties are based on gross concrete properties. The structure is divided into frames. Each equivalent frame is made up of the line of columns with the slab attached; the slab width is determined by lines of zero shear. If the structure is braced then not only are the column sizes minimized but the calculations are simplified as only vertical load needs to be considered on our equivalent frame. The equivalent frame is as follows (see Fig. 6.1) :

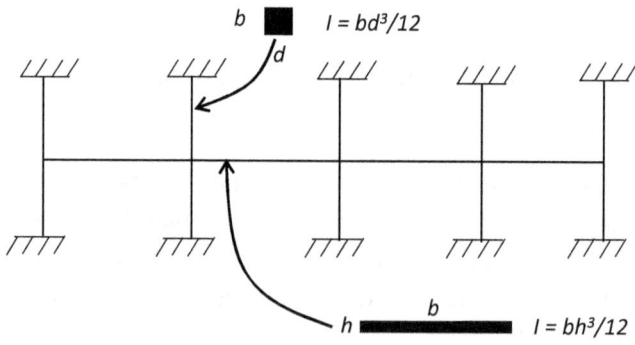

Fig. 6.1: Equivalent Frame

In reality the upper columns are rarely present at the time the floor is stressed. However this refinement (i.e. including only the lower column when considering the slab s/wt and prestress) is frequently ignored in practice (see Hurst 1998).

6.1.2.1. Modelling the slab/column connection.

It is well known that the slab/column junction is not 100% stiff. TR43:2005 allows this to be modelled by carrying out the analysis using a fictitious column length kl_{act} where $k > 1$, and defines $k = 0.5$ x (column spacing)/(column width + 6 x depth of slab).

So, for example,

Grid 8 m x 8 m; Slab 250 mm; Column 400 mm x 400 mm;

Then $k = 0.5(8)/(0.4+6 \times 0.25) = 2.1$

This modification may be ignored, if a more conservative answer is required (i.e. if ignored the moment at the junction between the slab and column will be higher than necessary).

6.1.3   Tendon Layout

Because of pre-compression in both directions slab behaves as a flat plate almost regardless of tendon arrangement. However, sufficient tendons must pass through the column zone to give adequate punching and progressive collapse resistance.

The minimum P_e/A needed to be effective > 0.7 N/mm^2 but at least 1.38 N/mm^2 is recommended if water tightness is critical (Lin 1981).

If the panel is square, a common arrangement is as follows:

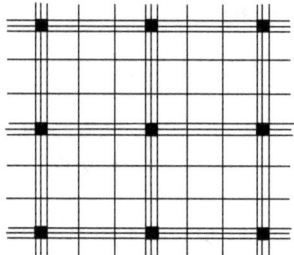

For square or irregular bays, a banded tendon layout can be used (less tendon crossing so maximizes drape):

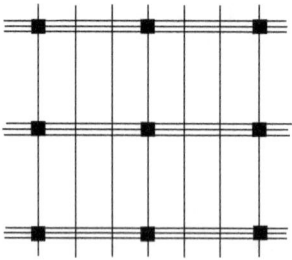

6.1.4. Allowable stresses for flat slabs

For flat slabs analyzed under *characteristic* loads using the equivalent frame method:

Location	Compression	Tension (with bonded reinforcement)
Support	$0.3f_{ck}$	$0.9f_{ctm}$
Span	$0.4f_{ck}$	$0.9f_{ctm}$

Note: for transfer use f_{ci} in place of f_{ck}; If no un-tensioned (i.e. conventional) reinforcement is provided, no tension is allowed at the support and $0.3f_{ctm}$ in the span.

The allowable stresses shown here are average stresses for the full panel width. In reality the hogging moments are sharply peaked around the columns: the moment at the column face is several times the moment midway between the columns. The sagging moment is much more evenly distributed.

If a F.E. (or grillage) analysis is used then the stresses given in the table above need <u>not</u> be used. Instead use the allowable stresses indicated in the following:

For flat slabs analyzed under *frequent* loads using FEA or Grillage:

Location	Compression	Tension (with bonded reinforcement)
Support	$0.4f_{ck}$	$1.2f_{ctm}$
Span	$0.4f_{ck}$	$1.2f_{ctm}$

Note: If no untensioned reinforcement is provided (or if tensile bars exceed 500 mm spacing), a tension of only $0.4f_{ctm}$ is allowed.

6.1.5. Vibration

As the floor construction is usually thinner than RC equivalent, this means its natural frequency is lower; thus the floor is more susceptible to vibration. In addition, PSC floors also contain fewer cracks so damping is less than RC. TR43 provides a detailed method whereby vibration effects can be checked.

TR43 requires *no vibration check* for normal office/residential loading if either,

- slab thickness > 250 mm with more than 4 panels (in any direction) or
- slab thickness > 200 mm with more than 8 panels (in any direction).

6.1.6.   Improving Economy of Post-tensioning

- The floor span should be a minimum of about 7 m.
- Where possible, reduce the length of the span of the end-frame for better distribution of bending moments.
- Where possible, introduce short cantilevers. This both improves the distribution of bending moment and, where the bonded system is used, eases the problem of placing the tendon through the column cage, as the anchor (around 300 mm wide) and the column reinforcement would otherwise clash.
- Reduce the column/wall stiffness (to reduce restraint).
- Always brace flat slab construction (e.g., by using walls) otherwise this construction, being heavy, is very prone to P-Delta effects (i.e. base moment magnification due to movement of c.g. of weight).

6.1.7.   Floor thicknesses

TR43 recommends that the span/thickness ratio should not exceed 42 for floors and 48 for roofs. It recommends the following thicknesses for post-tensioned spans up to 13 m.

Floor Types	ADL^1+LL (kN/m^2)	Span/Depth
Flat Plate[2]	2.5	40
	5.0	36
	10.0	30
Flat Slab with drops[2,3]	2.5	44
	5.0	40
	10.0	34
One-way slab with wide beam[4]	2.5	45/25
	5.0	40/22
	10.0	35/18
One-way slab with narrow beam[5]	2.5	42/18
	5.0	38/16
	10.0	34/13

Notes:

1. "ADL" = Additional Dead Load.

2. Square panel assumed.

3. Drop panel length ≥ span/3, depth ≥ 3h/4 where h is slab thickness.

4. 'Wide' Beam: typical width of beam = span/5

5. 'Narrow' Beam: typical width of beam = span/15

Note: To accommodate the *anchorage plates* (and provide a cover of 20 mm to each end) floor thickness should be a minimum of about 200 mm (bonded system).

6.1.8. Deflections of flat slabs

According to TR43 it is acceptable to base deflection calculation on the ELASTIC UNCRACKED deflections from an FEA/grillage analysis using the quasi-permanent loadcase and **modified accordingly to allow for creep and cracking**.

Loading	Factor applied to short-term elastic deflection value
Dead	3.0
Post-tensioning (effective)	3.0
Live	1.5

6.1.9. Additional anti-crack reinforcement

Provided in support region of flat slabs to improve crack control. The following is required by TR43.

- Placed as near to surface as cover limits allow.

- Reinforcement at least 0.075% of gross concrete cross section.

- Placed within $1.5h$ from column boundary.

- Bars at least $0.4L$ long.

- Spacing not greater than 300 mm.

Typically a grid of H16 bars is provided.

Note: No such un-tensioned reinforcement is required by TR43 in one-way systems (However minimum distribution reinforcement is still required in one-way slabs). Nevertheless, to maintain at least some hogging moment capacity away from the support (the tendon will be at mid-depth at around $0.1L$), some additional conventional reinforcement is suggested.

6.1.10   Other Reinforcement

At slab edges U-bars are placed. At least two longitudinal lacer bars are also provided in top and bottom.

Reinforcement is also required in the triangular unstressed areas between each anchorage. In particular, the slab around the corner column is often not prestressed if the tendons are bonded (Warner, 1998).

Bursting reinforcement is required and is calculated using section 8 of EC2 (see Chapter 8). This reinforcement usually takes the form of vertical one legged links in slabs and closed links in beams.

In flat plates punching reinforcement is usually provided around columns. Often prefabricated stud reinforcement is used in preference to conventional links in order to ensure full anchorage of punching reinforcement and speed up steel fixing. Punching is considered in more detail in the next section.

6.1.10.   Punching

There is some uncertainty in assessment of punching strength of a prestressed slab. Recommendations in various published reports disagree. For preliminary design it is suggested that tendons are simply ignored and punching resistance calculated using anti-crack reinforcement alone (i.e. as RC). This approach is conservative.

112

Note that solid slabs of span > 10 m are likely to require drop panels in order to provide sufficient resistance to punching.

6.1.11. One-way systems: Water resistance

The best way to achieve water resistance of slab is to prestress it both ways. Thus if a beam and slab system is used it is necessary to provide an extra tendon or two in the slab <u>parallel</u> to the beams.

6.2. SUMMARY OF PRELIMINARY DESIGN

- Choose section from section 6.1.7.

- Decide % of gravity load to be balanced => w_{bal}

- Choose the largest drape possible given cover limits (see relevant Tables in EC2), i.e. δ

- Calculate P_{eff} needed to achieve w_{bal} from

  $P_{eff} = w_{bal}L^2/8\delta$

- **Double value** of P_{eff} (i.e. assume jack to 70% F_{pk} and 30% losses)

- Use tendon area with this breaking load (see section 3.1.4).

- Check vibration.

6.2.1. Example: Preliminary design of post-tensioned flat plate office.

A floor slab consists of a flat plate of concrete of f_{ck} = 40 N/mm^2. Columns are on 8 m x 8 m grid forming three panels in the transverse direction and five panels in the longitudinal direction. Mild internal exposure; Fire resistance requirement 1.5 hrs. ADL = 1 kN/m^2 and LL = 4 kN/m^2. A plan is shown below. Estimate the slab depth and amount of prestressing tendons required for service conditions in bending. Tendons are bonded.

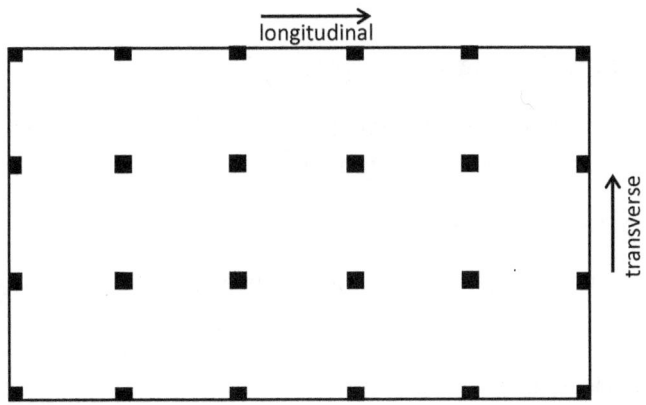

Solution

Assume entire self-weight is to be balanced.

Depth of slab:

From section 6.1.7. (Total imposed load = 5 kN/m^2) = 8,000/36 = 225 mm

Cover required:

Durability: EC2 part 1.1 clause 4.4.1.2 requires at least 45 mm cover to the side of the duct. Conventional reinforcement requires a cover of 30 mm.

Fire resistance: EC2 part 1.2 clause 5.2(5) requires an axis distance of 40 mm.

Drape available:

Check typical end span (controls).

Interior support: Anti-crack mesh of H16 assumed; tendon top under uppermost H16.

End: tendon at mid height.

(Use flat duct system i.e. duct of 19 mm x 75 mm; ignore actual likely position of strand *within* duct.)

Thus in Transverse Direction:

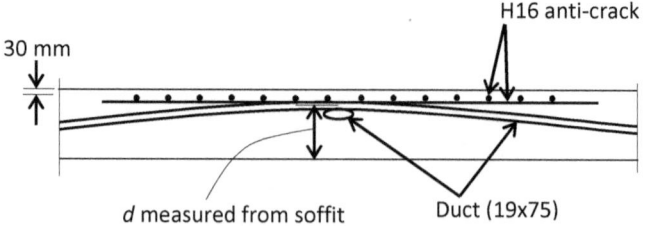

d measured from soffit   Duct (19x75)

Consider upper tendon:

At internal support: e_s = 225/2 – 30 – 16 – 19/2 = 57 mm.

Midspan: (No bars) => e_m = 225/2 - 45 - 19/2 = 58 mm.

End: e_e = 0.

Thus drape available, δ = 57/2 + 58 = 86 mm

Estimation of prestress force:

Load to be balanced, w_{bal} = 100% s/wt slab = 0.225x25 = 5.63 kN/m^2; Hence need P_{eff} = $w_{bal}L^2/8\delta$

where L is the distance between the points of inflexion of the tendon (roughly 0.9 x span).

Thus P_{eff} = 5.63x(0.9x8)^2/(8x0.9x0.085) = 476 kN/m.

Hence need F_{pk} = 2x476 = 953 kN/m

(F_{pk} = 2P_{eff} obtained by assuming stressing to 70% breaking and 30% losses).

From section 3.1.4. say use 15.2 mm "drawn" strands with max characteristic force = 300 kN each =>

Need 953/300 = 3.18 strands per meter;

4 strands per tendon => tendons at 4/3.18 = 1,250 mm centres.

Maximum spacing allowed = max centres 8 x slab thickness = 8x0.225 = 1.8 m > 1.25 m).

Check P_{eff}/A is reasonable = 476/225 = 2.1 N/mm² close to 2 N/mm² => reasonable.

Check vibration:

Section 6.1.5. => 225 mm slab requires min 6 panels. Have 15 panels => ok.

Use 4no. 15.2 mm "drawn" strands at 1250 mm centres.

6.3. CHECKLIST WHEN EXAMINING PT "SHOP" DRAWINGS PREPARED BY CONTRACTOR

- Ensure design for correct gravity loading; (the loads should appear on a drawing so the contractor can design the props);

- Identify type of tendon;

- Spot checks on ultimate moment capacity;

- Check punching shear strength;

- Ensure links are provided in any beam;

- Check other minimum reinforcement has been provided (anti-crack reinforcement in tops of flat slabs, edge, bursting, trimming, triangular unstressed regions).

- Check no reinforcement obstructs stressing pockets. Check pocket size and ensure it is to be filled with **non-shrink** mortar once grouting complete.

- Ensure enough room for anchorages if they are to be placed where the reinforcement is congested.

- Ensure detailing includes provision for post-punching resistance of flat plates (see later).

CHAPTER 7: PRE-TENSIONED ELEMENTS

This section gives the requirements for precast elements. Typically there are five stages that should be checked for most precast members:

1. Transfer;

2. Lifting from the mould;

3. Pitching (Horizontally-cast members 'pitched' to their vertical orientation. Refers mainly to columns and walls.)

4. Service;

5. Ultimate.

The procedures for checking transfer, service and ultimate have been already been covered.

7.1. LIFTING FROM THE MOULD.

A 50% suction and impact allowance is added to the self-weight of the member as it is removed from the mould. The element's ultimate strength is checked using this load case. The lifting points are positioned about $0.2L$ from the ends of the member (thus $M_{hog} \approx M_{sag}$).

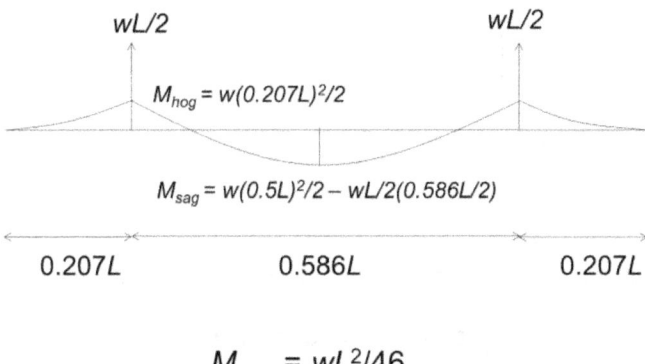

$wL/2$ $wL/2$

$M_{hog} = w(0.207L)^2/2$

$M_{sag} = w(0.5L)^2/2 - wL/2(0.586L/2)$

0.207L 0.586L 0.207L

$$M_{max} = wL^2/46$$

Lifting is usually done 15-18 hours after casting. The concrete cube strength is taken as 15 N/mm² (typically a strength of at least 20 N/mm² must be obtained from 2 match-cured cubes.) The characteristic cube strength at 28 days is usually around 60 N/mm².

7.1.1. Example

Check the lifting requirement for a 12 m long 300 mm square column which is lifted from the mould using lifting points at $L/5$ from each end (2.4 m). The proposed reinforcement is 4H20 bars with 8 mm diameter links and 30 mm cover. Assume concrete C12/15.

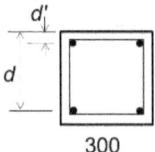

300

Solution: Self weight = 0.3x0.3x25= 2.25 kN/m

Allow 50% for suction and impact => 3.38 kN/m

Ultimate Moment M_{Ed} = 1.35x3.38x2.4^2/2 = 13.1 kNm

Shear V_{Ed} = 1.35x3.38x2.4 = 11.0 kN

d = 300-30-8-10 = 252 mm

Section analysis: 2H20 => T = 2x314x0.87x500 = 273 kN

$z \approx d - d'$ = 252 – 48 = 204 mm => M_R = Tz = 273x0.204 = 55.7 kNm > 13.1 kNm. Hence flexure okay.

Shear capacity without links

V_{Rdc} = bd(0.12(1 + (200/d)$^{0.5}$)(100$\rho_l f_{ck}$)$^{0.33}$)

=300x252x0.12(1+200/252)$^{0.5}$)(12x100x628/300x252)$^{0.33}$

= 300x252x0.48 = 36.2 kN > 11 kN => shear okay

7.2.   PITCHING

Pitching is achieved by lifting the element at about 0.3L from the top. A 25%-50% impact allowance is added to the self-weight of the element as it is pitched on site. Here the concrete strength is usually equal to the design strength (i.e. 28-day).

$$0.045wL^2$$

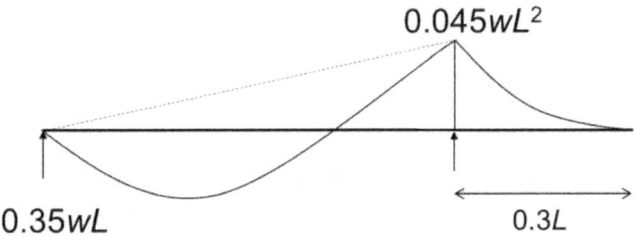

0.35wL 0.3L

The worst situation is just as element leaves the ground. For static load $M_{max} = 0.045wL^2$ (i.e. $wL^2/22$).

When lifting elements such as columns, it must also be ensured that the element does not swivel when lifted so the lifting point is located at the centre.

7.3.   HYBRID AND COMPOSITE CONSTRUCTION

"Hybrid" here refers to concrete cast at two different times acting together. This usually means a combination of precast and cast-in-place concrete. The two elements join to form a single *composite* member.

The composite member usually consists of a lower component (usually precast beam or slab) and an upper component (usually cast *in-situ* slab). So the term 'hybrid' is used to describe the mix of materials; 'composite' describes the interaction of the materials.

Prestressed composite construction is either:

1. "Un-propped"; or

2. "Propped".

The terms refer to the absence or presence of propping to the lower beam/slab during the addition of the *in-situ* topping. Un-propped construction is more common.

7.3.1.   Un-propped construction

It consists of two stages (see Fig. 7.1):

> – Stage 1. Precast beam placed.

Thus beam carries own self-weight.

> – Stage 2. *In-situ* slab placed.

The precast beam carries the weight of the *in-situ* slab/topping. Loads added after the *in-situ* slab hardens, i.e., *ADL* and *LL*, are carried by composite section.

Stage 1: precast beam placed:

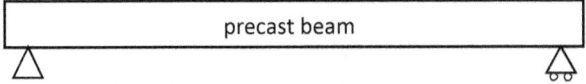

Stage 2: *insitu* topping placed:

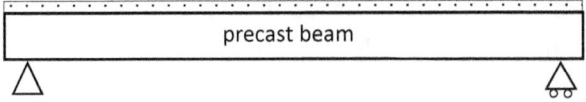

Fig. 7.1: un-propped construction

7.3.1.1. Critical load stages in <u>Un-propped</u> construction:

1. Transfer: Prestressing force P_i; self-weight of precast beam w_o;
2. During casting of in-situ concrete: Prestressing force $\sim P_e$; self-weight w_o; weight of *in-situ* concrete;
3. Service: Prestressing force P_e ; beam self-weight w_o; weight of *in-situ* slab. Composite section carries any subsequent load (w_{ADL} and w_{LL}).

7.3.2. Propped construction

There are three stages (see Fig. 7.2):

— Stage 1. Precast beam placed.

- Stage 2. Prop is added at one or more places along length.

- Stage 3. *In-situ* slab placed. Propping released after *in-situ* slab has hardened.

The loads added subsequently, i.e., *ADL* and *LL*, are carried by <u>composite</u> section.

Stage 1: precast beam placed:

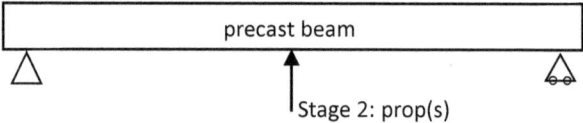

Stage 3: *insitu* topping placed; allowed to harden then prop(s) removed:

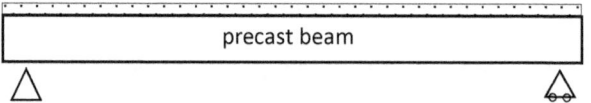

Fig. 7.2: propped construction

7.3.2.1.   Critical load stages in <u>Propped</u> construction:

1. Transfer: Prestressing force P_i; self-weight of precast beam w_o.

2. During casting of *in-situ* concrete (beam propped): P_e and weight of *in-situ* concrete on <u>propped</u> beam.

3. Service: propped beam carries weight of *in-situ* slab. These stresses are said to be "trapped" by construction method. <u>Composite</u> section carries any subsequent load (w_{ADL} and w_{LL}).

7.3.3.   Advantages of composite construction:

- The benefit of precasting (standardized sections, re-use of forms, improved quality control and finish) can be availed of;
- Formwork and scaffolding are largely eliminated for the *in-situ* slab leading to rapid site construction;
- There is very little interference to the work below.

7.3.4.   Stresses in Composite Members:

Stresses in the *in-situ* concrete are normally low. Thus, lower strength concrete can be used; this means we use concrete of a lower elastic modulus E_c. The analysis of stresses on the composite section is thus performed on the <u>transformed</u> section. So for calculations the **width** of the *in-situ* slab is be *reduced* by $E_{in\text{-}situ}/E_{PC}$, i.e., by the modular ratio.

7.3.4.1.   Example 1 – Un-propped Construction

A simply supported composite floor spans 12 m. It is made from precast prestressed beams placed at 1000 mm crs and a top slab of *in-situ* concrete 150 mm thick. Precast beams are 600 mm deep and 200 mm wide and are prestressed with a straight tendon 150 mm below the centroid. Effective prestress force P_e = 650 kN. Precast f_{ck} = 50 N/mm² and *in-situ* f_{ck} 25 N/mm².

Assuming that E_c of *in-situ* is 0.8 of E_c of precast, calculate stresses in concrete at midspan under an imposed load of 5 kN/m². Beams are un-propped during pouring and hardening of *in-situ* concrete. Neglect the weight of formwork between beams.

Properties of Precast section: A_{pc} = 120,000 mm², I_{pc} = 3.6 x 10⁹ mm⁴, Z_1 = Z_2 = 12 x 10⁶ mm³

Solution

Locations:

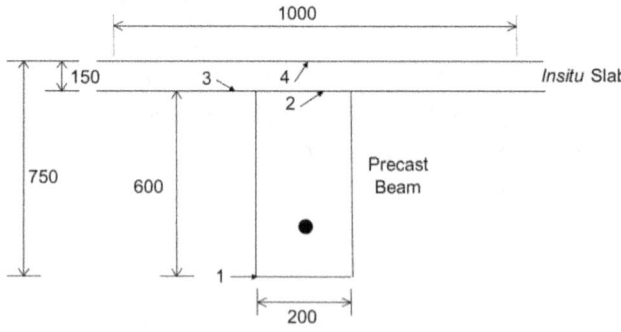

We will calculate stresses at locations 1, 2, 3 and 4.

Precast Beam: Stresses due to prestress alone:

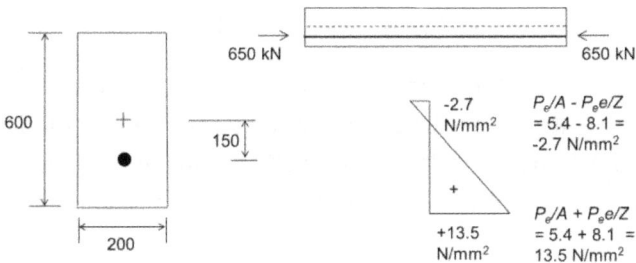

Un-propped construction:

The precast prestressed beam carries:

1.   Self-weight, $w_O = A_{pc} \times 25 = 120,000\ (10)^{-6} \times 25$

= 3.0 kN/m

2.   In-situ slab, $w_D = A_{slab} \times 25 = 1.0 \times 0.15 \times 25$

= 3.75 kN/m

=> Total load, $w_{TOT} = w_O + w_D = 6.75$ kN/m

Midspan moment due to these loads is:

$M_{TOT} = w_{TOT} L^2 /8 = (6.75)(12)^2/8 = +122$ kNm

- Stresses due to prestressing only:

    $f_1 = 13.5$ N/mm^2 (bottom fibre)

    $f_2 = -2.71$ N/mm^2 (top fibre)

- Stresses due to w_{TOT} only, are:

$$f_1 = M_{O+D}/Z_1 = 122/12 = -10.2 \text{ N/mm}^2$$

$$f_2 = M_{O+D}/Z_2 = 122/12 = 10.2 \text{ N/mm}^2$$

Imposed load 5 kN/m^2 applied after *in-situ* slab hardens:

$$M_{LL} = w_{LL} L^2/8 = (5)(1.0)(12)^2/8 = +90 \text{ kNm.}$$

Stresses under this load are computed based on composite action. Stresses in the composite section are proportional to ratio of the elastic moduli. If E_c of in-situ concrete is 0.8 of E_c of precast beam, the modular ratio n = 0.8.

To calculate properties of composite section, first find its centroid:

$$A_{comp} = A_{pc} + A_{slab}$$

$$= 120{,}000 + 0.8 \ (1000 \times 150) = 240{,}000 \text{ mm}^2$$

Distance from c.g. pc to c.g. comp is y:

$$y = [A_{pc}x0 + A_{slab}x(300+150/2)]/A_{comp} = [0 + 120{,}000(300 + 75)]/240{,}000 = 188 \text{ mm}$$

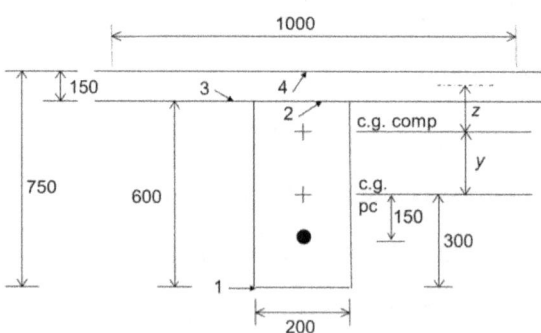

Now use the parallel axes theorem to find the second moment of area about the c.g. comp axis.

$I_{comp} = I_{pc} + A_{pc} y^2 + I_{slab} + A_{slab} z^2$

$= 3.6 \times 10^9 + 120{,}000 \times 188^2 + (1/12)(0.8 \times 1{,}000) \times 150^3 + (0.8 \times 1{,}000) \times 150 \times (300 + 150/2 - 188)^2 = 12.2 \times 10^9$ mm^4

$Z_{1comp} = I_{comp}/y_1 = 12.2 \times 10^9/(300+188) = 25 \times 10^6$ mm^3

$Z_{2comp} = I_{comp}/y_2 = 12.2 \times 10^9/(300-188) = 109 \times 10^6$ mm^3

$Z_{3comp} = I_{comp}/y_3 = 12.2 \times 10^9/(300-188) = 109 \times 10^6$ mm^3

$Z_{4comp} = I_{comp}/y_4 = 12.2 \times 10^9/(300+150-188) = 46.8 \times 10^6$ mm^3

Additional stresses due to $M_{LL} = 90$ kNm can now be computed.

$f_1 = M_{LL}/Z_{1comp} = 90/25 = -3.6$ N/mm^2

$f_2 = M_{LL}/Z_{2comp} = 90/109 = 0.8$ N/mm^2

Stresses in *in-situ* slab are:

$f_3 = (M_{LL}/Z_{3comp}) \times n = (90/109) \times 0.8 = 0.7$ N/mm^2

$f_4 = (M_{LL}/Z_{4comp}) \times n = (90/46.8) \times 0.8 = 1.5$ N/mm^2

Note: stresses in in-situ slab have to be multiplied by modular ratio to obtain actual/correct stresses.

Stresses:

•　　Tension:

–　　Total tensile stress in precast = 0.3 N/mm^2.

–　　Allowable tension = f_{ctm} = 4.1 N/mm^2.

•　　Compression:

–　　Total compression stress in precast = 8.3 N/mm^2.

–　　Allowable compression = $0.45f_{ck}$ = 22.5 N/mm^2.

–　　Total compression stress in *in-situ* = 1.5 N/mm^2.

–　　Allowable compression = $0.45f_{ck}$ = 11.3 N/mm^2.

Summary:

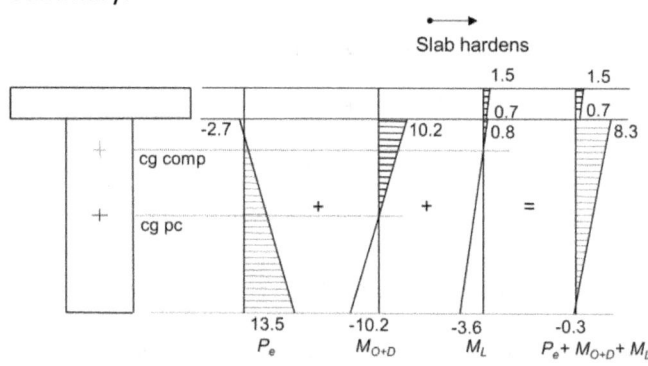

Note: If there was no composite action between the precast and *in-situ* the maximum stresses in the precast would be:

f_{top} = -2.7+10.2+7.5 = 15.0 N/mm^2;

f_{bot} = 13.3-10.2-7.5 = -4.2 N/mm².

7.3.4.2. *Example 2-Propped Construction*

Repeat the previous example but this time during the pouring and hardening of the *in-situ* concrete the precast beam is propped by a single prop at midspan.

Solution

The only change in the calculations comes from the application of the *in-situ* slab. Its weight is applied to propped structure. (Beam self-weight is of course applied to the 12 m simply supported beam).

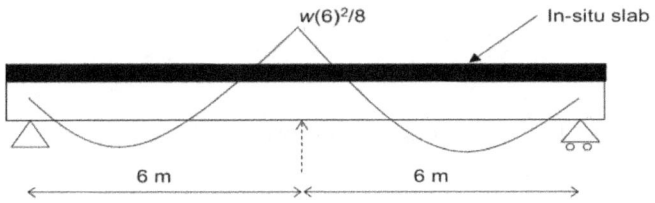

Stresses at midspan

Self-weight: w_O = 3.0 kN/m;

$M_O = w_O L^2/8 = 3.0(12)^2/8 = +54$ kNm

In-situ slab 3.75 kN/m = w_D

$M_D = -w_D(L/2)^2/8 = -3.75(6)^2/8 = -16.9$ kNm

Moment at midspan M_{O+D} = 54-16.9 = 37.1 kNm

Once *in-situ* slab is placed the stresses in precast are:

$f_1 = M_{O+D}/Z_1$ = 37.1/12 = -3.1 N/mm²

$f_2 = M_{O+D}/Z_2$ = 37.1/12 = 3.1 N/mm²

Summary: stresses (N/mm²)

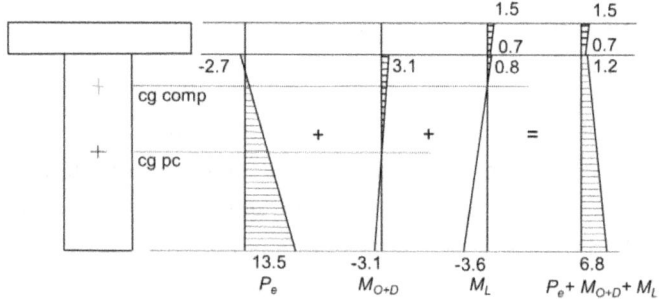

7.3.5. Ultimate Strength of Composite Section

Procedure for finding M_u of composite section is the **same** as for regular section, i.e., assume that the construction procedure has no influence on M_u. The same assumption is made for the calculation of vertical shear capacity.

7.3.6.   Interface Shear Transfer

In order for composite action to take place, the <u>interface</u> <u>shear</u> must be transmitted across the interface (junction) between the precast beam and the *in-situ* slab. This is shown in Fig. 7.3.

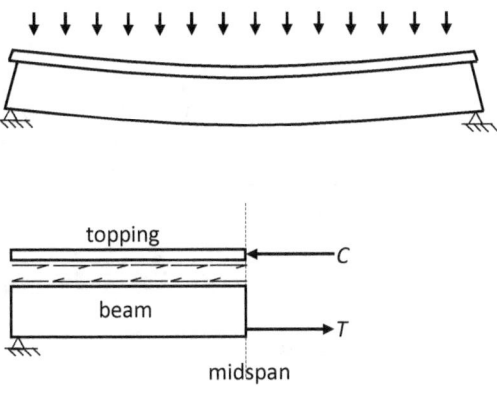

Fig. 7.3: interface shear

For precast sections with a large top surface (e.g., T, TT, hollow core, planks), the top surface can normally be left 'as-cast'. This normally provides a sufficiently effective mechanism for shear transfer between the *in-situ* topping and the precast elements. For these sections no reinforcement across junction is needed in most cases. However, for sections with relatively narrow top surfaces (e.g., I-beams), since the area of contact is relatively narrow, some links to resist interface shear may be needed. In most cases, normal vertical links required for "beam" shear are merely extended to reinforce the junction.

7.3.6.1.   Design of Shear Connectors:

Consider a U.D.L. The shear force diagram is as shown in Fig. 7.4.

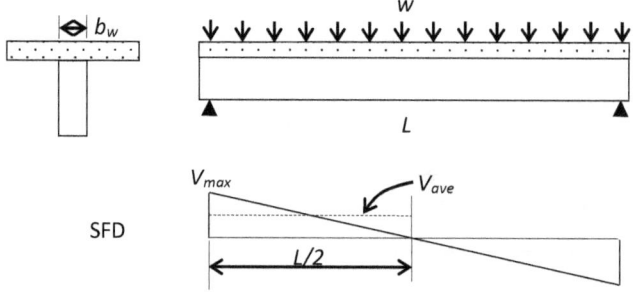

Fig. 7.4: shear force diagram for UDL

The shear **force** diagram thus shows the distribution of vertical shear **stress** along span of beam. As long as the beam remains elastic then the distribution of **vertical shear** and so **horizontal shear**, is according to the shear force diagram. Of course, at ultimate a concrete beam is clearly <u>not</u> elastic. However it is <u>assumed</u> that the *distribution* of interface shear (i.e., v_h) is similar to that of the shear force diagram.

The interface shear force to be transferred is the total force at ultimate in the slab (conservatively assuming the entire slab thickness is in compression):

$F_H = b_{eff}\, h_f\, (0.57\, f_{ck})$

b_{eff} = effective width of slab, h_f = thickness of slab and f_{ck} is the characteristic cylinder strength. See Fig. 7.5.

The effective width of a flange, b_{eff}, should be based on distance, l_o, between points of zero moments:

$b_{eff} = b_w + b_{eff,1} + b_{eff,2}$

where

$b_{eff,1} = (0.2b_1 + 0.1l_o)$ but $\leq 0.2l_o$ and $\leq b_1$

$b_{eff,2} = (0.2b_2 + 0.1l_o)$ but $\leq 0.2l_o$ and $\leq b_2$

b_w width of web (if any)

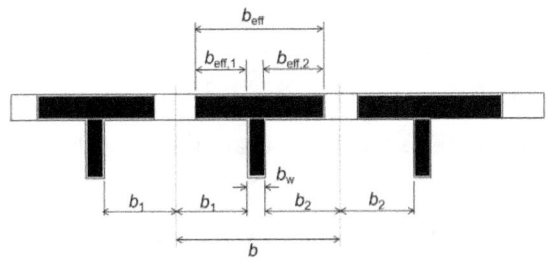

Fig. 7.5: Effective width of flange

Thus the shear stress is $v_{h,ave} = F_H/(b_w L/2)$

Now this average shear stress is *distributed* according to the elastic shear force diagram. So, in the case of a *U.D.L.*, the maximum shear stress $v_{h,max}$ is at the support and $v_{h,max} = 2v_{h,ave}$

Allowable interface shear stress (EC2 cl. 6.2.5)

$$v_{Rdi} = cf_{ctd} + \rho f_{yd}\mu \le 0.5 \upsilon f_{cd}$$

where,

$f_{ctd} = f_{ctk}/1.5 \; ; f_{cd} = 0.85 f_{ck}/1.5$

$c = 0.4$ and $\mu = 0.7$ for 'roughened surfaces' (i.e. 3 mm at 40 mm),

ρ = Area of steel crossing interface and properly anchored both sides/Interface Area

$\upsilon = 0.6(1 - f_{ck}/250)$

7.3.6.1.1. Example

Simply supported span 12 m. Loaded by a *U.D.L.* Section shown below.

Check that provision of shear connectors is adequate.

$f_{ck, in\text{-}situ} = 30 \text{ N/mm}^2$ and $f_{ck, precast} = 50 \text{ N/mm}^2$

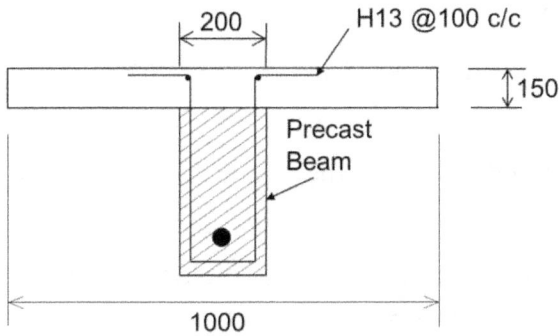

Solution

Assume full thickness of slab in compression at ultimate (conservative assumption).

$b_{eff,1} = b_{eff,1} = b_1 = 400$ mm => $b_{eff} = 1,000$ mm

The total horizontal force in the slab at midspan at ultimate is $F_H = 0.57 f_{ck} b_{eff} h_f = 0.57(30)(1000)(150)$

= 2,565 kN.

The average horizontal shear stress at the interface is $v_{h,avg} = 2F_H/b_w L = 2 \times 2,565 \times 1000/200 \times 12,000$

= 2.13 N/mm²

Load is uniformly distributed, therefore $v_{h,max} = 2v_{h,ave}$

$v_{h,max} = 2(2.13) = 4.26$ N/mm²

Allowable interface shear

$v_{Rdi} = 0.53 + \rho f_{yd}\mu \leq 0.5 \upsilon f_{cd} = 4.5$ N/mm² where,

f_{ck} = 30 N/mm^2 => f_{ctd} = 2/1.5 = 1.33 N/mm^2; f_{cd} = 0.85x30/1.5 = 17 N/mm^2

c = 0.4 and μ = 0.7 for 'roughened surfaces' (i.e. 3 mm at 40 mm),

ρ = Area of steel crossing interface and properly anchored both sides/Interface Area

$\upsilon = 0.6(1 - f_{ck}/250) = 0.53$

Maximum shear allowed = 4.5 N/mm^2 > 4.26 N/mm^2

Thus we reinforce the interface s.th 0.53 + $\rho f_{yd}\mu$ = 4.26 N/mm^2

Now μ = 0.7 and f_{yd} = 500/1.15 = 435 N/mm^2

Thus ρ = (4.26-0.53)/(0.7x435) = 0.012

Thus A_h = 0.012x200x1000 = 2,450 mm^2 per metre.

i.e. required area of shear connector per meter run is 2,450 mm^2.

Try 2 legs H13 at 100 mm centres.

A_h = (1000/100)x(133x2) = 2,660 mm^2 > 2,450 mm^2 => ok

Note: As mentioned previously, the required area A_h can very often be provided by projecting the precast beam links into the cast in-situ concrete slab. The links then become full-height of the composite beam. In fact since only full-height links are fully effective in the composite beam, it is recommended that in *all* cases the links are projected and properly anchored. Otherwise the composite beam has a deficiency in shear, as shear cracks approach the top of the section at the ULS.

7.4. END-SHEAR

7.4.1. Full-height beam

In beam-like members (i.e. "isolated") the ends are often dapped to save headroom. We have to be especially careful with the detail at the ends in these cases.

But even if the full-height beam is supported we still need to add reinforcement in an isolated member. A truss is formed at the end. Well-detailed reinforcement allows this truss to form. A strut C_1 forms in the concrete; tie T_1 is needed for equilibrium of this strut. See Fig. 7.6.

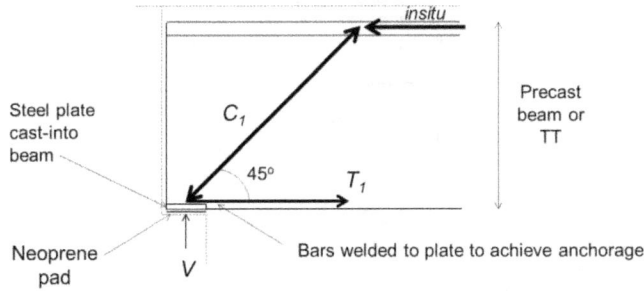

Fig. 7.6: Full-height beam support

Tie T_1

Vertical equilibrium at node: $C_1 \sin 45° = V$

Horizontal equilibrium at node: $C_1 \cos 45° = T_1$

=> $\tan 45° = V/T_1$ i.e. $T_1 = V/\tan45°$

In addition, a horizontal friction force F_t (typically = 0.4V) is added to T_1. Pre-tensioning strands cannot provide any of this resistance as they are not well anchored at the support region. Thus conventional reinforcement should be added in all cases when the member is 'isolated'.

7.4.2. Members with Pockets

If a pocket is used at the support the following figure 7.7 shows models of behaviour that are appropriate in cases (a) where the pocket is more than the height/3 and (b) where it is no more than height/3. The worked example of the long span floor includes case (b).

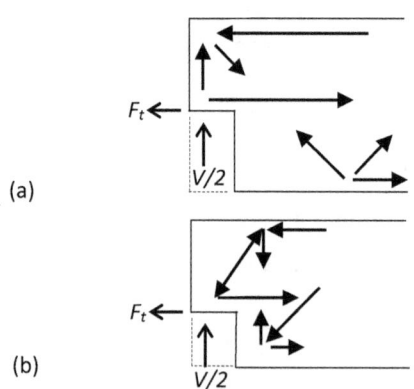

Fig. 7.7: Models for beam support (a) pocket > beam depth/3 and (b) pocket < beam depth/3 (Elliott 2013)

141

CHAPTER 8: END-BLOCKS/ANCHORAGE ZONES

8.1. POST-TENSIONING

When a concentrated compressive force is applied to a concrete prism tensile stresses are generated at 90 degrees to this compression. These stresses are known as 'bursting stresses'. By Saint-Venant's 'principle', the compressive stresses become uniform at about the member depth away from the end.

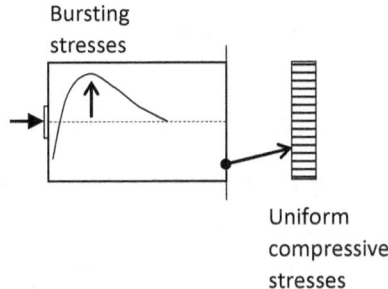

Fig. 8.1: bursting stresses

EC2 recommends the use of a strut-and-tie model, with 34 degree struts, to design the end-block. See Fig. 8.2.

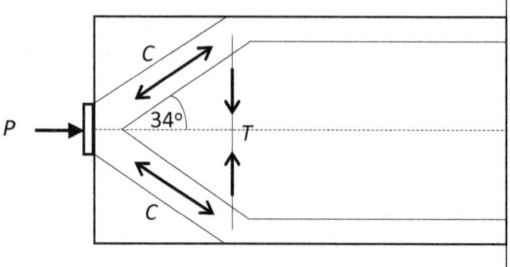

Fig. 8.2: strut-and-tie model for end-block design

Thus tie force $T = 0.5P \tan34° = 0.34P$

And strut force $C = 0.5P/\cos34° = 0.6P$

In addition, the compressive stresses at the bearing plate are to be limited to

$$f_{Rdu} = 0.67f_{ck}(A_{c1}/A_{c0})^{0.5} \leq 2f_{ck}$$

where A_{c0} is the loaded area of the bearing plate and A_{c1} is the maximum area of the same shape as A_{c0} which can be inscribed in the total area A_c. This is shown below in Fig. 8.3.

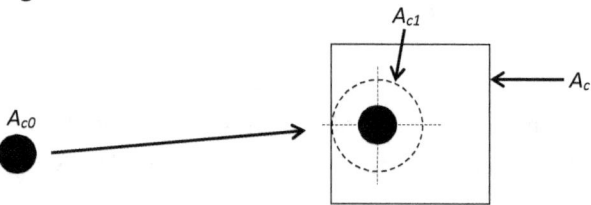

Fig. 8.3: parameters for design

The compressive stresses in the strut should not exceed $0.4(1-f_{ck}/250)f_{ck}$. If the reinforcement stress is limited to 300 N/mm² then no checks on crack widths are needed. The reinforcement should be spread over a length equal to the member height. A load factor $\gamma_p = 1.2$ is applied to jacking force P.

8.1.1. Example

(Mosley 2007)

A prestressing force P of 1000 kN is applied to the end of a member 400 mm x 500 mm using 4 conical anchorages as shown below.

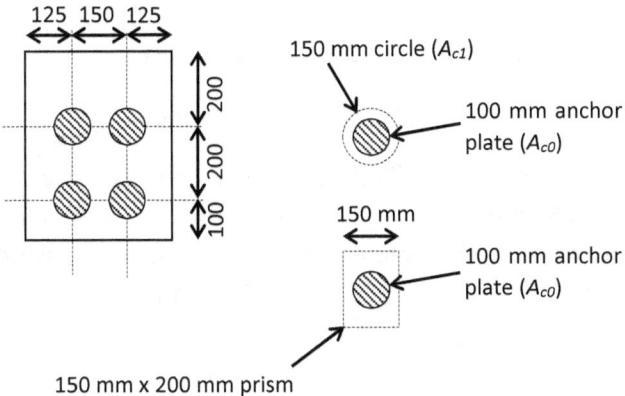

150 mm x 200 mm prism

Solution

Check bearing stress = $(\gamma_p \times P)$/loaded area

= $1.2 \times 250 \times 10^3/(\pi 100^2/4) = 38.2$ N/mm^2

Allowable bearing stress = $0.67 f_{ck}(A_{c1}/A_{co})^{0.5} = 0.67 \times 40$ [π $\times 150^2/ \pi \times 100^2]^{0.5} = 40.2$ N/mm^2 > 38.2 N/mm^2 => ok

Check stress in strut:

Allowable compressive stress = $0.4(1-f_{ck}/250)f_{ck} = 0.4(1-40/250)40 = 13.44$ N/mm^2

Actual stress in strut = Force/Area = $0.6 \times 1.2 \times 250 \times 10^3/(200 \times 150 \times \cos 34°) = 7.21$ N/mm^2 => ok

Reinforcement

Force in tie = $0.34P = 0.34 \times 1.2 \times 250 = 102$ kN

Area required, $A_s = 102 \times 10^3/300 = 340$ mm^2

Use 3H10 closed links (78.5x6 legs = 471 mm^2) at, say, 50 mm, 125 mm and 200 mm from the end face (arrangement of anchors implies equivalent prisms around each anchor of 150 mm x 200 mm.)

Check combined anchorage

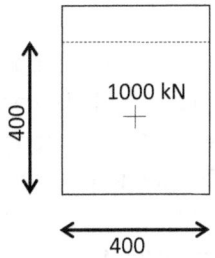

The effect of the combined anchorage can be considered as a 1000 kN force applied to a 400 mm x 400 mm face.

Tensile force = 0.34x1.2xP = 408 kN

Area required, A_s = 408x10^3/300 = 1360 mm^2

Use 6H12 closed links (6x2x113 = 1356 mm^2) distributed over a length of 400 mm

Details of reinforcement:

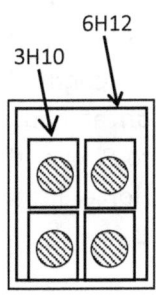

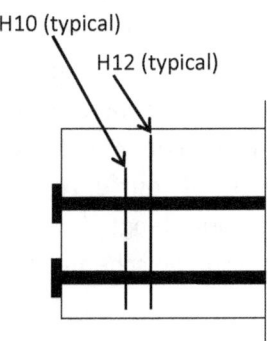

8.1.2. Reducing congestion of steel

Sometimes the design requires so many links in the end-block region that severe congestion results. It is very important that the concrete in the end-block is dense and well compacted (Abeles 1981). Congestion is often due to the excessive conservatism of the code; for example, the code only considers links spread to about the member depth as being effective. However experiments have shown that the code method is conservative and links placed up to twice this distance are effective (Ibell 1992).

8.2. PRE-TENSIONING

There is no anchorage hardware as, just like with a conventional reinforcing bar, the bond between the tendons and the concrete is relied upon to anchor the tendon. When concrete gains sufficient strength the tendons are cut. The force in the tendon at end of member is, of course, zero and increases to P_e over the "transmission length". The build-up of force in the tendon is shown schematically below (Fig. 8.4).

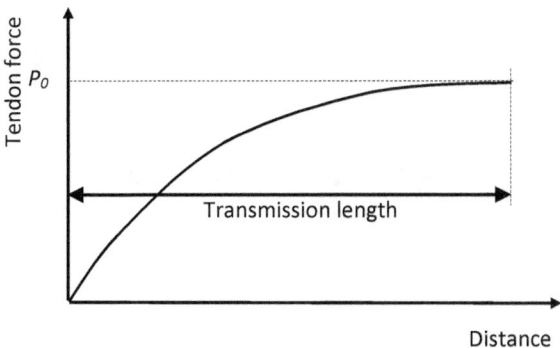

Fig. 8.4: transmission length in pre-tensioned members

For design, the typical transmission length, l_{pt} (="development length") is normally taken as about 70 dia. for strands and 100 dia. for wires/bars. Thus for strands transmission length is usually of the order of 900 mm while for wires it is usually of the order of 1500 mm.

The actual transmission length depends on many factors, for example the following: Concrete strength (particularly degree of compaction); the surface condition of steel; cross sectional profile of steel; diameter of steel and the "wedging" effect at the cut end due to the tendency of the steel to revert to its original diameter (Fig. 8.5).

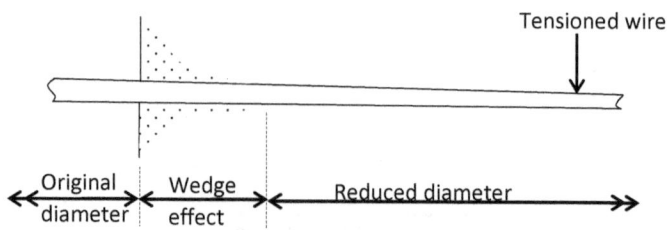

Fig. 8.5: wedge effect

As its effectiveness is very dependent on bond, prestressing steel must be kept well-spaced apart. The recommended spacing is about 3-4 diameters (Lin 1981).

Although the prestressing force is applied over a longer length than post-tensioned steel, it is still essential that there is no splitting of the concrete. In larger beams it is usual to provide stirrups around pre-tensioned steel especially when there is steel near bottom and tops of section. One simple design method is that specified by ACI:318, namely that the area of vertical stirrups A_t is suggested to be $0.021P_ih/f_sl_t$ mm^2.

where

P_i is initial prestress force (in N),

h is the total member depth (in mm),

f_s is the allowable stress in the stirrups (about 140 N/mm^2)

l_t is transmission length (in mm).

The ACI code suggests this reinforcement, A_t, should be uniformly distributed over the end $h/5$ of the beam.

Note: In hollow core slab units full reliance is placed on the tensile strength of the concrete and no reinforcing is used. This simplifies manufacturing, allowing extrusion/slip forming to be used.

Despite the addition of reinforcement, anchorage zones may still fail. This might be because curing near slab edge is generally not as good as away from edges, or when wet concrete placed often concrete near edge is more watery, thus at risk of honeycombing. Match cured test cubes should be placed at edges for more reliable strength information.

CHAPTER 9: LEARNING FROM FAILURES

9.1. PROGRESSIVE COLLAPSE OF FLAT SLAB/PLATE STRUCTURES

If a collapse is not prevented from growing from a *local* collapse into a *global* collapse the result is called a "progressive collapse".

Now, concrete flat slabs are vulnerable to punching shear failure. This is a local failure. Careful detailing ensures that they are *not* vulnerable to progressive collapse too. Once punching shear failure has occurred (for whatever reason) the reaction can no longer be transferred to the column, so is transferred to neighbouring columns. This may result in failure occurring there too. Thus, some or all of the floor can fall. Clearly we must prevent it falling.

There have been several examples of the punching failure of flat slabs using conventional RC leading to progressive collapse, e.g. Sampoong department store, Seoul, 1995; Cocoa Beach condominium, Florida, 1981; Bailey's Crossroads condominium, Virginia, 1973. These and similar cases are described in various books, for example, MacAlevey (2010).

9.1.1. Unbonded

If unbonded tendons are used, it is recommended (for instance by the first edition of TR43 i.e. TR43:1994 and of recent editions of ACI:318), that at least two tendons pass through column cage to increase resistance to progressive collapse in the event of punching shear failure at column. These tendons act as catenaries (i.e., cables) to allow slab to stay suspended once the slab has failed in punching.

As mentioned previously, in the unbonded system each strand is a tendon and is individually anchored. Thus the anchor is small. Thus it is easily accommodated in column cage.

9.1.2. Bonded

In the case of the bonded system however, strands are grouped in 4s or 5s so the anchorage is large (typically around 300 mm). In addition, the duct is large (the 'flat duct' is often about 75 mm wide) and must remain reasonably straight over its length. Thus it is very difficult to accommodate the anchorage within the column. Thus to get progressive collapse resistance we have to do one of the following:

1. Ensure any bonded tendons are placed no further than half of the slab depth from the column (TR43, 2005) as shown in Fig. 9.1.

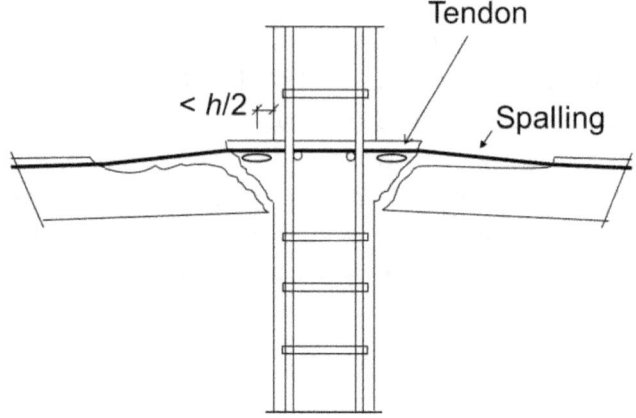

Fig. 9.1: tendon close to column

2.   Alternatively, use cantilevers to avoid the clash of anchor and column reinforcement so the tendon can pass through the column, or

3.   Use unstressed <u>bottom</u> bar reinforcement (required by EC2 for conventional RC flat slabs). Place this reinforcement within the column reinforcement cage as shown in Fig. 9.2.

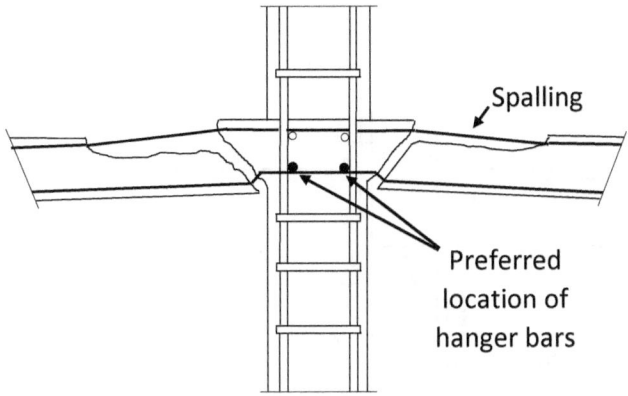

Fig 9.2: unstressed reinforcement (bottom bars)

9.1.3. Design

If the tendons are able to pass through column cage, then no calculations are necessary. If bottom bars are used, then there should be a minimum of four bars (Fig. 9.3). For design it should be assumed the bar is fully stressed i.e. yield strength is reached (Hawkins 1979).

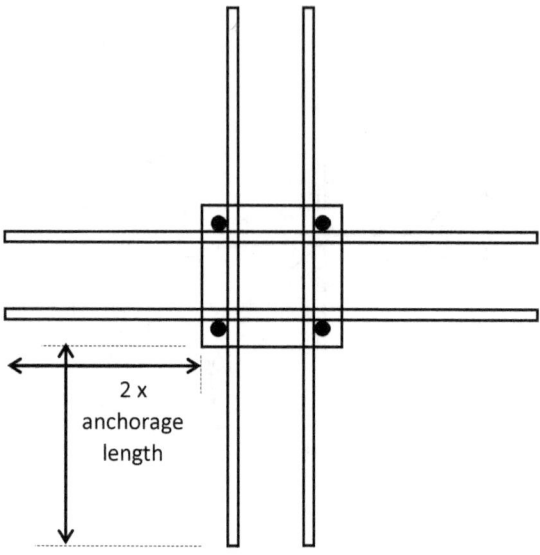

Fig. 9.3: layout of bottom bars

9.1.2.1. Example of provision of bottom bars

A flat plate structure spans 7.2 m in each direction, has a slab thickness 200 mm of f_{ck} = 30 N/mm². The structure is residential. Assume LL = 2 kN/m² and finishes of 1 kN/m². The internal column is 600 mm x 600 mm. design the bottom reinforcement for a typical internal column. Use EC safety factors.

Solution:

Accident Load = DL + 0.5 LL = (0.2x25+1+0.5x2) = 7 kN/m^2
=> reaction = 7x7.2x7.2 = 363 kN

Capacity of H16 = $A_s f_y$ = 201x500 = 100.5 kN

Thus no. of H16s = 363/100.5 = 3.6

Provide 4H16 bars.

EC2 requires that the anchorage length for a 16 mm bar in this situation is about 35 diameters. So in our case a typical bar is 4x35x16+600 = 2840 mm say 2900 mm.

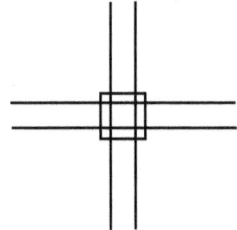

9.2. RESTRAINT

Horizontal movement of floor is due to elastic shortening, creep, shrinkage and temperature changes. Allow for a total long-term movement of at least 0.6 mm per metre of floor (use this to estimate Δ in the following). An RC floor tends to move about 60% of this. Movement of floor should be allowed otherwise restraint forces develop.

It is reasonable to *ignore* restraint in the design if **all** of these are true (TR43 2005):

- If the average P/A is less than about 2 N/mm^2,
- If the floor not very long (i.e. < 50m),
- If there are no walls or "large" columns (i.e. of column dimension > about 600 mm) restraining movement.

The following figure (Fig. 9.4) shows a floor supported on columns and a wall. Both the immediate and long-term movement is towards the wall. The columns, being much more flexible, are pulled over, bending into an 'S' shape in the process.

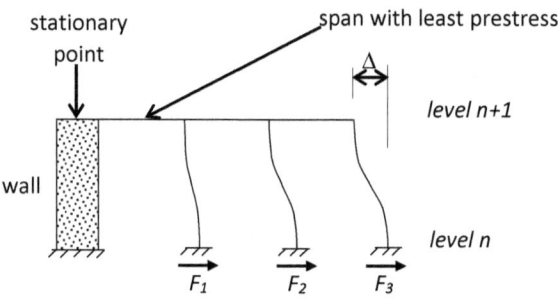

Fig. 9.4: restraint of floor

At the perimeter each column is forced to deflect an amount Δ. The force in each column can be calculated assuming double curvature in the column (Fig. 9.5).

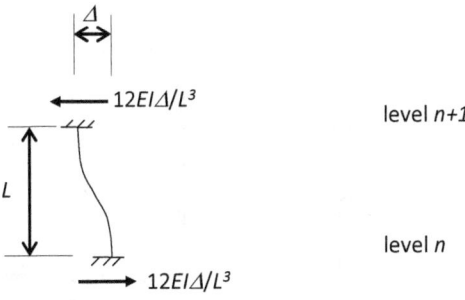

Fig. 9.5: force implied by deflected shape

Where L is the height of the column; E is the modulus of the concrete; EI for the column may be reduced due to creep and cracking in the column. A reduction of 50% is reasonable.

Thus the total tension in the floor = F_1+ F_2+ F_3.

If this significantly (say more than 10%) reduces P_e/A then consider a temporary release.

The building layout plan on the left (Fig. 9.6(a)) is a good one from the point of view of restraint (a single fixed point—the core--means all the columns are pulled towards it) while that on the right is a poor one (four equally stiff restraints mean movement must occur towards each corner—'X' shaped cracking is likely).

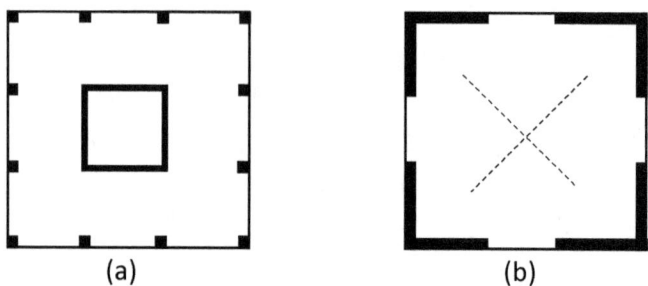

Fig. 9.6: layouts resulting in more or less restraint

The following plan (Fig. 9.7) would be a good layout from the point of view of restraint. The perimeter walls have little resistance to being pulled over. Hence movement is reasonably free to occur towards the core.

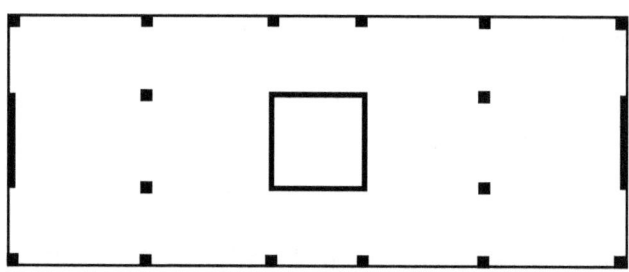

Fig. 9.7: layout resulting in less restraint

Having two stiff cores at each end of the building is also a poor layout (Fig. 9.8).

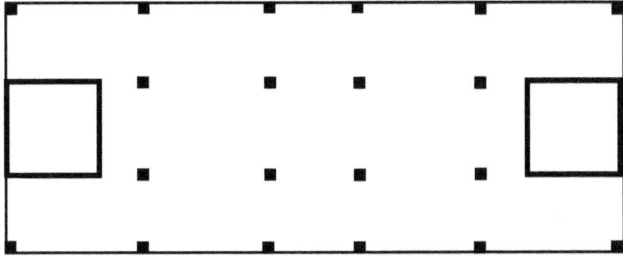

Fig. 9.8: layout resulting in greater restraint

On a multi-storey building, restraint drops as height increases. Post-tensioning is easier for *suspended* slabs compared to *ground bearing* slabs as the restraint is lower if the slab is suspended. To reduce restraint for a ground bearing slab, use leveled sand blinding under the slab along with two layers of plastic sheet on top of the blinding.

9.2.1. Excessive Restraint

The act of prestressing requires that the member be allowed to move relatively freely. If such movement is *not* allowed the following consequences should be expected:

- Prestressing P/A goes to restraints!
- Effect of tendon eccentricity (and thus any upward load) will get into member.

- Cracks likely
- Ultimate capacity is not affected.

9.2.2. Restraint Cracks:

Although frequently observed in post-tensioned slabs, these cracks are often cosmetic; occasionally they need careful investigation (e.g. if in a zone of high shear). For durability, fill those cracks > 0.25 mm wide with epoxy as late as possible in construction.

There is a tendency to over-react to these cracks and specify permanent slip details. Remember that once the structure is cracked the restraint is partially or totally relieved. Permanent slip-details should be only introduced as a last resort as they reduce redundancy and bracing to columns as well as being troublesome to construct and therefore expensive.

9.2.3. Allowing movement:

There are several ways of allowing movement and so reducing restraint:

- Provide RC "pour strips" (Fig 9.9) at least 1 m wide (for stressing access) about 45 m apart or next to large restraints if there is a need to isolate them. Allow these to remain open for at least 28 days (about 40% of creep and shrinkage would have occurred by then).

- Introduce releases, e.g., temporary slip bearings; make some walls non-structural (e.g. masonry).

- Alter the sequence of construction so restraining elements are built later.

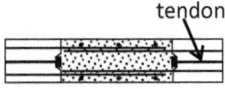

Fig. 9.9: RC pour strip

In some cases the restraint cannot be eliminated by any of the above methods so additional reinforcement should be added. A frequent case is along a long concrete wall cast monolithically with the slab (Fig. 9.10).

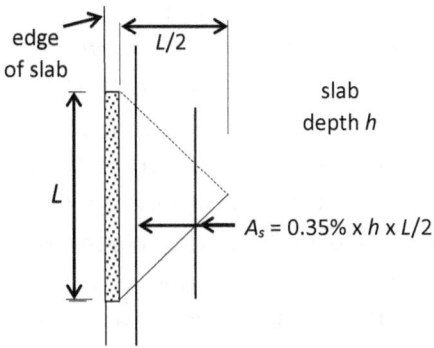

Fig. 9.10: additional reinforcement parallel to wall.

Little prestress gets into the slab within the triangle shown, as the slab cannot freely compress near the wall. TR43 requires this reinforcement.

9.3. DURABILITY

Once cast inside concrete steel is protected from corrosion by the high pH environment cement paste provides. Unfortunately this protection can be lost over time. Carbon dioxide or Sulphur dioxide gases seep into the concrete, especially in cracked regions, and reduce the pH. However, there is another cause of deterioration: chlorides seep into the concrete, usually in water, and attack the steel directly. Prestressing helps by reducing cracking.

In most cases any deterioration can be repaired well before it becomes too late and the weakened structure collapses. Warning signs are rust stains and cracks caused by rusting steel expanding and perhaps even sagging as the steel is lost. Unfortunately, in the case of Gwas Bridge, no such warning signs were given. See for example MacAlevey (2010) or Woodward (1988). This case led to a ban on grouted post-tensioning in UK *bridges* in the 1990s which was later lifted when specifications for grouting were tightened (Tilly 2002). In buildings, such a failure seems much less likely. However, car park slabs in cold climates are at risk from chloride attack from road-salt brought in on the tyres of vehicles.

9.4. MISCELLANEOUS CASES

9.4.1.   Case 1: Anchorages and Restraint

(Thornton 1993)

The first structure we'll consider is a car park with post-tensioned beams and one-way slabs and a conventionally reinforced vertical structure. The columns were circular. The problem was this: how to anchor these tendons without affecting the appearance of the joint?

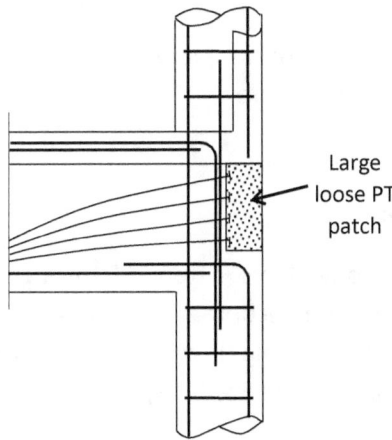

Fig. 9.11: connection between PT beam and circular column

The solution adopted was that beam tendon anchorages were set at <u>column</u> mid-depth (Fig. 9.11).

Problems:

1. The anchorage plates created plane of weakness within the column.
2. A large patch was used to cover the anchorage. It was unreinforced (as is usual) and the concrete was not non-shrink so it was not secure.
3. The reinforcement from the column and beam had to be bent down too early, creating a weak beam-column joint.

Result:

Severe deterioration at every beam-column joint. Early demolition of structure.

Avoidance:

The beam could have extended through the column. Individual tendon anchorages and plugs could have been used. Where possible fewer, stronger tendons could have been used. (See Fig. 9.12)

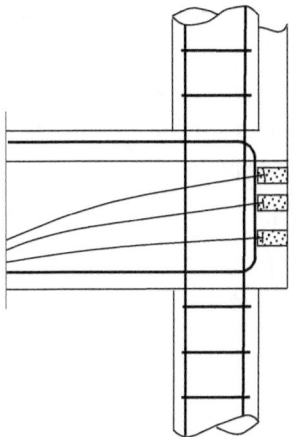

Fig. 9.12: modified connection detail

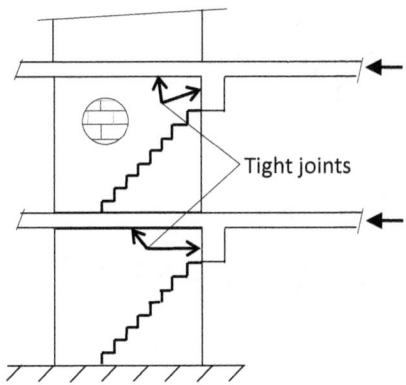

Fig. 9.13: cracking of masonry walls

At the same parking structure additional problems were encountered relating to masonry walls (Fig. 9.13). The masonry enclosure for a stair was built tightly to the floor, thus preventing the floor from moving. When the floor was prestressed cracks appeared that followed the wall joints. The lesson here is simply to provide soft joints so that some movement is allowed.

9.4.2.   Case 2: Missing Dead Load

(Thornton 1993)

This case concerns a transfer beam in an office structure. Normally the procedure would to add prestress in increments as the structure above is completed. However, in this case there appears to have been a breakdown in communication between the designer and the contractor. This resulted in the full prestress being applied to the beam before the structure above had been completed. (See Fig. 9.14).

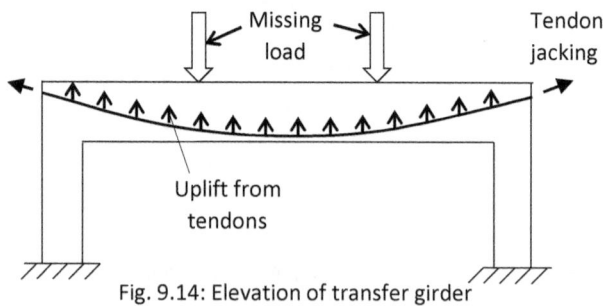

Fig. 9.14: Elevation of transfer girder

Not surprisingly, the beam hogged upwards, the concrete in the compression zone crushed, and the beam failed.

In addition, the detailing was poor: many tendons of same profile created a plane of weakness (Fig. 9.15). Thus cracks following the path of the tendon could be seen (The rule of thumb when detailing conventional RC is that the cumulative width of the steel is limited to no more than 40% of the section; this can be used for PSC detailing too. Of course spaces should be left at regular intervals for a vibrator).

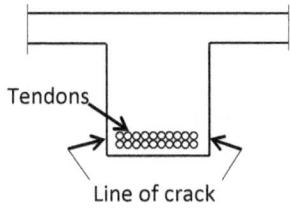

Fig. 9.15: poor detailing of transfer beam PT

A similar case of a prestressed raft slab is presented by Kaminetzky (1991).

9.4.3. Case 3: Force Patterns During Construction.

(Thornton 1993)

This case also concerns a five storey parking structure: here, a second-floor column was damaged just below level 3 as a result of unanticipated moments during placing of concrete at level 4. The configuration during construction was less favourable than the completed structure. See Fig. 9.16.

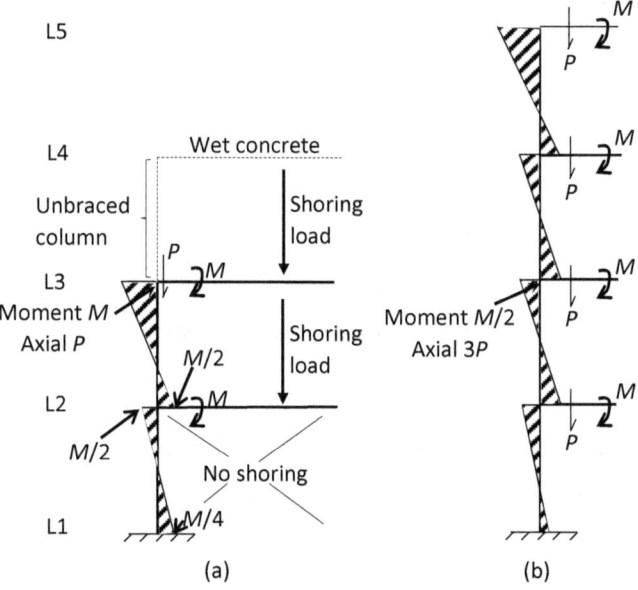

Fig. 9.16: (a) during construction; (b) as assumed in design

The structure was designed as if fully completed (see (b)). Thus it was assumed that all levels were fully braced.

However, during construction there was no bracing at level 4 (clearly no horizontal force can pass through wet concrete!). This means any bending moment M applied at level 3 could only be transferred downwards. In addition, the column strength was lower as the full axial load was not acting and the concrete itself was young. This resulted in the formation of a hinge just below level 3.

Essentially, the problem arose because the final design live load was much lower than the load imposed by construction. This is a common problem with parking structures and residential structures as the design live load is so low (typically 2.5 and 1.5 kN/m^2 respectively).

On such low-rise structures consider shoring to ground level throughout construction; this means that the structure carries little load during construction.

9.4.4. Other problems in car parks

High aspect ratio columns (known as 'blade' columns), are common in car parks. They are used to improve circulation and sometimes also to ensure the column qualifies as a 'wall' for fire resistance reasons (aspect ratio > 4). However, they often cause restraint problems. Consideration should be given to allowing the column to be squarer, say by increasing the concrete strength.

9.4.5. Case 4: Segmental Construction

(Kaminetzky 1991)

An ice hockey rink roof consisted of three 45 m span post-tensioned girders each made from three pre-tensioned beams, post-tensioned after erection (Fig. 9.17).

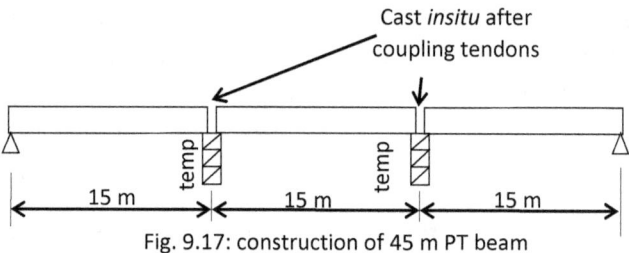

Fig. 9.17: construction of 45 m PT beam

Details of the beam at the coupling points are shown below (Fig. 9.18).

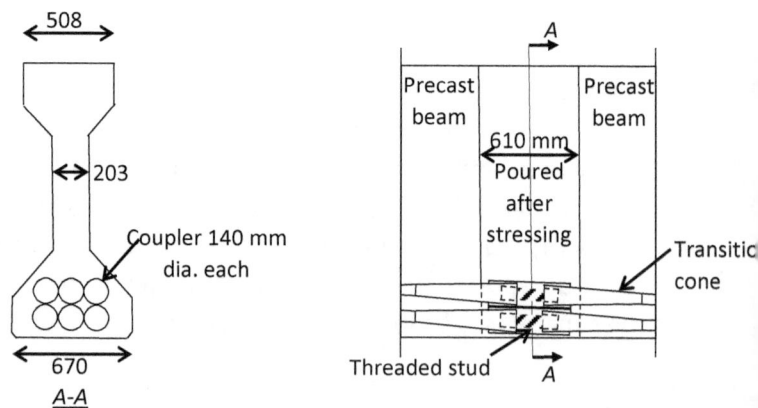

Fig. 9.18: coupling points

Post-tensioning was completed on the first and second girders. The third was being stressed when it hogged and failed. Later the other two girders also collapsed. Two factors were blamed for the failure:

- There was no allowance for movement of couplers. In addition, concrete had entered the ducts either side of the *in-situ* splice.
- Couplers were large (140 mm in diameter) and no allowance was made in design for their presence. The result was a large overstressing of bottom flange.

Lessons:

1. The duct should be continuous around the tendon splice, so that the coupler can move during stressing. Otherwise there can be no elongation of the tendon and so no prestressing.
2. Make provision in design for the absence of concrete at couplers, as couplers are often large.
3. Always compare the jack pressure gauge reading with actual tendon elongation reading. Doing this would have detected that only 15 m (i.e., 30%) of the tendon was being stressed.

CHAPTER 10: DESIGN EXAMPLES

10.1. FLAT PLATE EXAMPLE TO TR43

The following is the layout for a floor of an office building. It consists of a flat plate utilizing bonded tendons. The storey height is typically 3.7m. The structure is braced using structural walls.

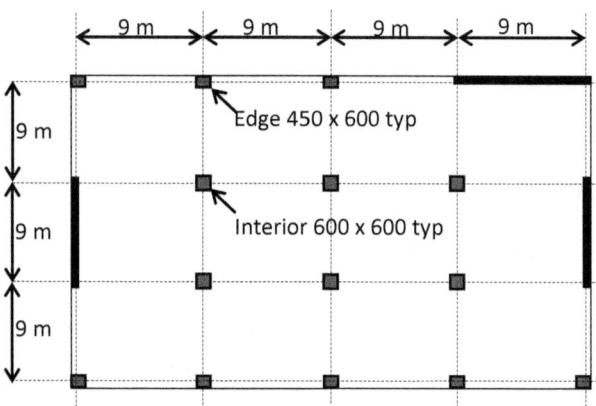

The following working loads are specified: live load of 3 kN/m²; superimposed load due to flooring and partitions assumed equivalent to a uniform load of 0.2 kN/m² and 1 kN/m² respectively, and a services load of 0.5 kN/m². The fire resistance requirement is 1 hour and the exposure condition is XC1 (="mild"). The concrete is normal weight (γ = 25 kN/m³) with f_{ck} = 35 N/mm² at 28 days and 25 N/mm² at transfer (3 days). Strands are "dyform" type; consisting of 7-wire steel of 15.2 mm nominal diameter and area A_{ps} = 165 mm^2 with a characteristic breaking load of F_{pk} = 300 kN each. E = 195 kN/mm^2. The strands are stressed from both ends (usually stress from both ends when tendon > 20 m) to $0.75F_{pk}$ and grouped in 4s or 5s in 19mm ducts.

1. Verify thickness of floor slab required.

2. Develop a preliminary layout of tendon/bar reinforcement based on an ideal tendon profile.

3. Carry out punching calculations for a typical internal column.

Use centreline moments for design for convenience (although EC2 allows moments calculated at the face of supports to be used).

Solution

Using Table of floor thicknesses recommended by TR43 in section 6.1.7. the recommended span/depth ratio is 36 (total IL = 3+1+0.2+0.5 = 4.7 kN/m^2) so try a 250 mm thick slab (9000/36=250). This slab has a self-weight of about 6.25 kN/m^2.

Consider a transverse bay (i.e., north-south).

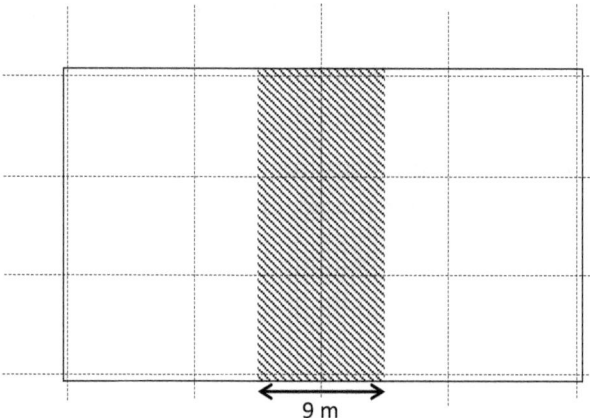

9 m

Tendon Profile

Cover to steel: 1 hr fire resistance and "mild" (XC1) exposure: Tables 4.4N and 4.5N of EC2 => 35 mm to ducts and 25 mm to reinforcement. Top anti-crack steel consists of H16s. Thus distance from duct centre-line to top and bottom surface of slab therefore 51 mm (= 25+16+19/2) and 45 mm (= 35+19/2) respectively. The actual position of the strand within the duct is ignored.

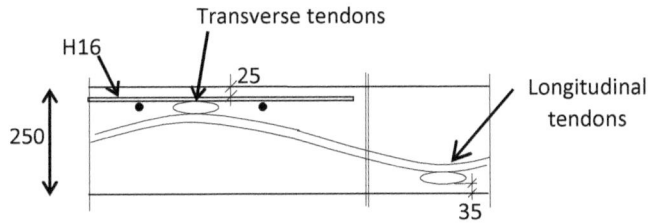

The following are assumed:

— Jacking force $P_j = 0.75F_{pk}$

— Initial force $P_i = 0.7F_{pk} = 0.7 \times 300 = 210$ kN (i.e., 5% losses)

— Effective $P_e = 0.56F_{pk} = 0.56 \times 300 = 169$ kN (i.e., total of 25% losses; $0.75 \times 0.75 = 0.56$).

Preliminary calculations are based on these assumed losses and idealized tendon profile.

Analysis is by the Equivalent Frame Method. We will use the actual (i.e. unmodified) column lengths.

Slab: $I = 9 \times 0.25^3/12 = 0.0117$ m^4
             $A = 9 \times 0.25 = 2.25$ m^2

Interior Col: $I = 0.6^4/12 = 0.0108$ m^4
             $A = 0.6 \times 0.6 = 0.36$ m^2

Edge Col: $I = 0.6 \times 0.45^3/12 = 0.00455$ m^4
             $A = 0.6 \times 0.45 = 0.27$ m^2

$E_c = 34{,}000{,}000$ kN/m^2

The equivalent frame model used for computer analysis is as follows:

176

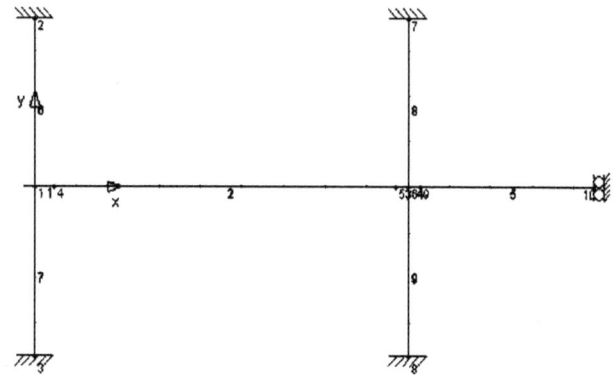

Balance <u>all</u> of slab self-weight, i.e., 6.25 kN/m^2.

Drape available in exterior span controls since it is least.

—   Span: e = 250/2-45 = 80 mm

—   Support: e = 250/2-51= 74 mm

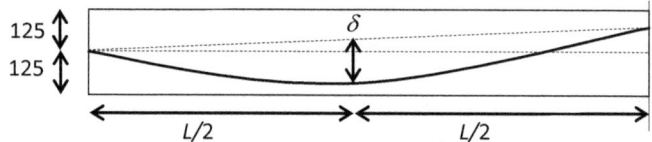

Thus drape δ is (80)+(74)/2 = 117 mm

Exterior span:

$P_e = w_{bal}L^2/8\delta = (6.25\text{x}9)\text{x}9^2/(8\text{x}0.117) = 4867$ kN

Number of strands is 4867/169 = **28** strands (7x4), where 169 kN is P_e of one strand.

Interior span:

Use same force, reduce drape to compensate, since want the same w_{bal}. Thus profile is as follows:

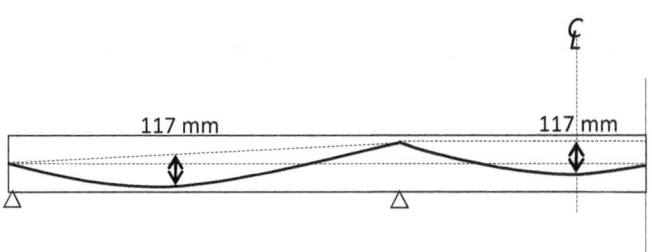

Effect of Prestress:

Due to the prestress <u>upward</u> load w_{bal} = 6.25 kN/m², i.e. 56.25 kN/m on each span.

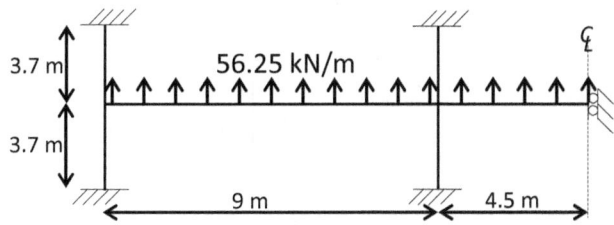

Analysis under this upward load gives the following bending moment diagram (kNm). These moments are the total moments M_{TOT}.

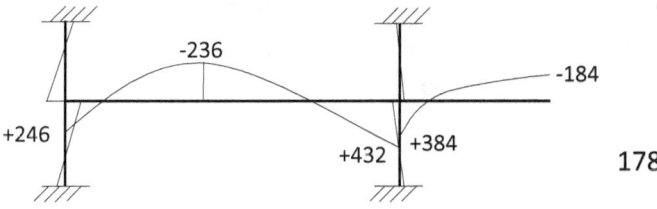

178

Primary moments are shown below.

Ext. mid-span : e = 80 mm;

Thus $P_e\,e_{max}$ = 4867x80 = 389 kNm;

Int. support e = 74 mm; $P_e\,e_{max}$ = 4867x74 = 360 kNm

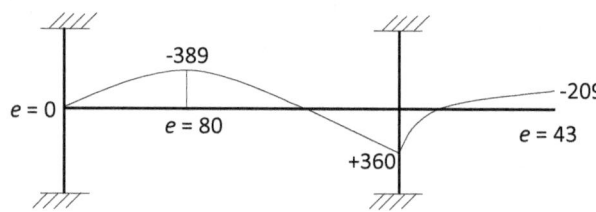

Secondary moments are shown below.

$$M_2 = M_{TOT} - M_1$$

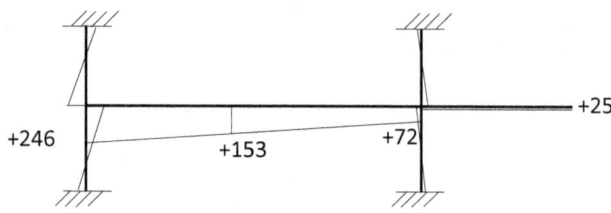

Note that the secondary moment in the perimeter column is large (123 kNm).

Analysis is now required under the remainder of load, i.e., the **unbalanced load** (Additional dead load, services, partitions and live load) = 4.7 kN/m^2

Additional dead load:

ADL (flooring) $w = 0.2$ kN/m², i.e., 1.8 kN/m on each span.

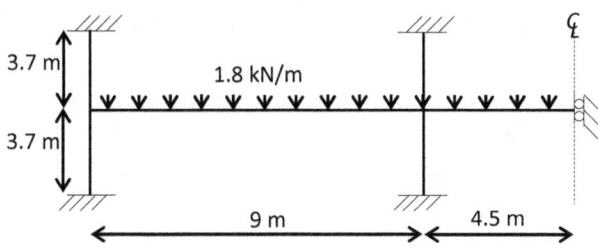

Live load case 1:

$w = 4.5$ kN/m², i.e., 40.5 kN/m on each span.

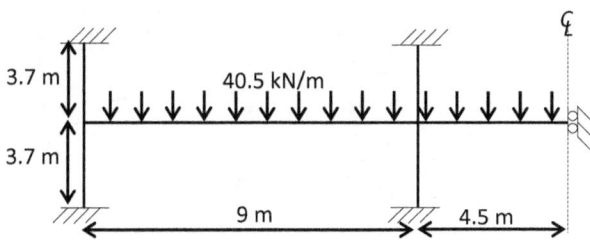

Live load case 2

$w = 4.5$ kN/m², i.e., 40.5 kN/m on exterior spans only.

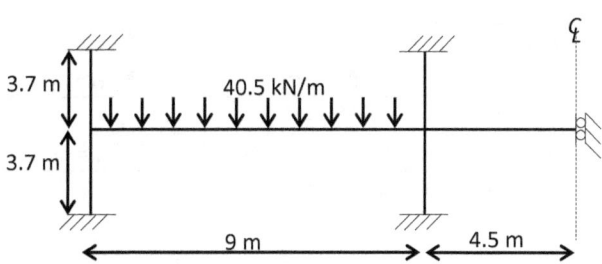

Live load case 3

$w = 4.5$ kN/m², i.e., 40.5 kN/m on interior span only.

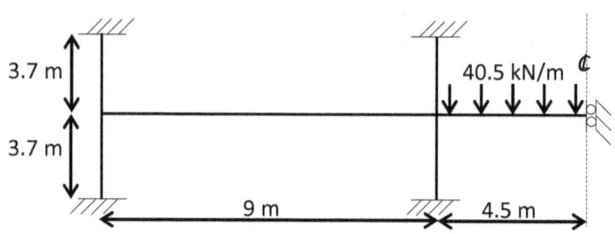

Slab Service Moments (1.0ADL+1.0LL)

Load	Exterior Span			Interior Span	
	Left Supp't	Mid-Span	Right Supp't	Left Supp't	Mid-Span
ADL	-8	+7.5	-14	-14	+6
LL1	-182	+166	-312	-277	+133
LL2	-196	+177	-271	-27	-27
LL3	+14	-12	-40	-250	+160
Design	-204	+184.5	-326	-291	+166

Allowable service stresses

For flat slabs analyzed using equivalent frame method:

Location	Compression (N/mm²)	Tension (N/mm²)
Support	$0.3f_{ck} = 10.5$	$0.9f_{ctm} = 2.88$
Span	$0.4f_{ck} = 14.0$	$0.3f_{ctm} = 0.96$

f_{ck} at service = 35 N/mm² so f_{ctm} = 3.2 N/mm²

Note: Bonded reinforcement is provided near the tension face in the support region but not in the span region.

Service Stresses

Exterior Span:

Left support; Worst case moment: 204 kNm (hogging)

$f_{top} = P_e/A + M_{unbal}/Z = (4867/9*250) -$
$204,000*6/(9*(250)^2) = +2.16 - 2.18 = - 0.02 \ N/mm^2$

$f_{bot} = +2.16 + 2.18 = + 4.34 \ N/mm^2$

These results are show in the first row of the following table.

Summary: (Moments in kNm and stresses in N/mm^2).

Location	M	P_e/A	M/Z	f_{top}	f_{bot}
Ext Span:					
Left Supp't	-204	2.16	-2.18	-0.02	+4.34
Mid-span	+185	2.16	2.0	4.16	+0.16
Right Supp't	-326	2.16	-3.48	-1.32	+5.63
Int Span:					
Left Supp't	-291	2.16	-3.1	-0.9	+5.26
Mid-Span	+166	2.16	1.77	3.93	-0.03

Transfer Stresses

P_e gives an *effective* M_{TOT} which is sufficient to balance entire *DL*. At transfer the prestressing force is P_i (giving an *initial* M_{TOT}) is $P_i = 0.7F_{pk}$ and $P_e = 0.56F_{pk}$. Thus factor applied to *effective* M_{TOT} moment diagram to get transfer moments (*initial* M_{TOT}) is 0.7/0.56 = 1.25

Slab s/wt. = "DL". Due to the self-weight of slab $w = 6.25$ kN/m^2, i.e., 56.25 kN/m on each span. We have already analyzed this case above but with the loads reversed.

Moments at Transfer (*DL+initial M_{TOT}*): (kNm)

	Exterior Span			Interior Span	
	Left Supp't	Mid-Span	Right Supp't	Left Supp't	Mid-Span
DL	-246	+236	-432	-384	+184
1.25xM_{TOT}	+307	-295	+540	+480	-230
Net	+61	-59	+108	+96	-46

Allowable stresses:

Location	Compression (N/mm^2)	Tension (without bonded reinf.) (N/mm^2)
Support	$0.3f_{ck} = 7.5$	$0.3f_{ctm} = 0.78$
Span	$0.4f_{ck} = 10.0$	$0.3f_{ctm} = 0.78$

f_{ck} at transfer = 25 N/mm^2 so f_{ctm} = 2.6 N/mm^2

Note: Bonded reinforcement is not provided near the tension face in either the support region or span region (net load is upwards)

Transfer Stresses

Exterior Span:

Left support; Worst case moment: 61 kNm   (sagging)

$f_{top} = P_i/A + M_{net}/Z = 5880/(9*250) + 61,000*6/(9*(250)^2) = +2.61+0.65 = + 3.26 \text{ N/mm}^2$

$f_{bot} = +2.61-0.65 = + 1.96 \text{ N/mm}^2 \text{ (comp => OK)}$

Summary: (Moments in kNm and stresses in N/mm^2).

Location	M	P_i/A	M/Z	f_{top}	f_{bot}
Ext Span:					
Left Supp't	61	2.61	0.65	**3.26**	**1.96**
Mid-span	-59	2.61	0.63	**1.98**	**3.24**
Right Supp't	108	2.61	1.15	**3.76**	**1.46**
Int Span:					
Left Supp't	96	2.61	0.98	**3.23**	**1.27**
Mid-Span	-46	2.61	0.49	**2.13**	**3.10**

Clearly the stresses are well within limits at both service and transfer. Thus we can proceed to tendon layout.

Transverse Direction:

- Drape 117 mm

- 28 strands per panel.

- To use 6 ducts group as: 4 ducts with 5 + 2 ducts with 4 strands ($\sum$ = 28 strands);

- Place all of the tendons/panel in a band centred on columns.

Longitudinal Direction:

- Tendon Eccentricity:

  - Span 80-19 = 61 mm

  - Support 74-19 = 55 mm

- Ext span: drape = 61 + 55/2 = 89 mm

- Strands/panel = 28x(117/89) = 37;

- Group into 8 ducts: 5x5 + 3x4 (=37);

- Distribute tendons uniformly.

Summary: tendon layout

All strands dyform nom dia 15.2 mm

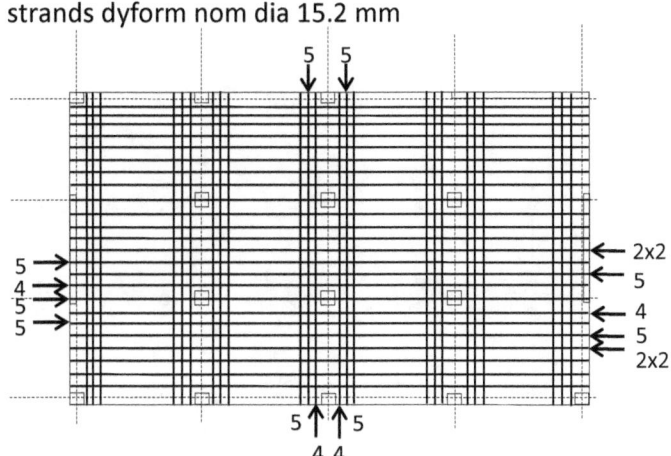

Additional anti-crack reinforcement

Minimum anti-crack reinforcement A_s = 0.00075x(concrete area).

9000 mm width => 0.00075x9000x250 = 1688 mm^2 => use 9H16 bars (A_s = 1809 mm^2).

Bars placed in width of ([3x(slab thickness) + column width]) = 1350 mm centered on the column.

Check bar spacing = 1350/8 = 170 mm < 300 mm OK

The length of the reinforcing bars is 0.4x9000 = 3600 mm.

Exterior column: 4800 mm width; => 0.00075x4800x250 = 900 mm^2 => use 5H16 bars (A_s = 1005 mm^2).

Summary: anti-crack reinforcement (top) layout

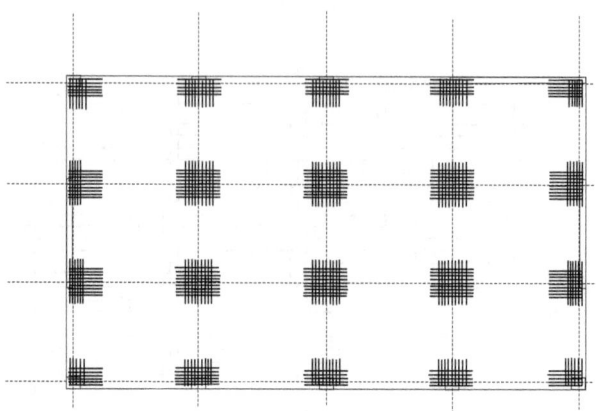

Slab Ultimate Moments

Moments due to (1.35DL+1.5LL) + worse of (1.1 and 0.9)M_2 are shown in the following table (kNm).

	Exterior span			Interior span	
	Left Supp't	Middle	Right Supp't	Left Supp't	Middle
DL	-246	+236	-432	-384	+184
ADL	-8	+7.5	-14	-14	+6
M_2	+246	+153	+72	+25	+25
$LL1$	-182	+166	-312	-277	+133
$LL2$	-196	+177	-271	-27	-27
$LL3$	+14	-12	-40	-250	+160
Ult BM	-416 (-208)*	+769 (+873)*	-1005	-930	+524

*after redistribution

For example: Exterior span: Left support: Elastic moment= 1.35x(DL + ADL) + 1.5xLL + 0.9xM_2 = 1.35(246+8) + 1.5x(196) – 0.9x(246) = 416 kNm (hogging)

After redistribution of 50% moment = 208 kNm (Note EC2 does not limit redistribution of edge column transfer moments.)

First Interior Support

Ultimate hogging capacity required is **1005 kNm** (see table).

Section is 9 m wide and reinforced with 7x4 = 28 strands and 9H16s.

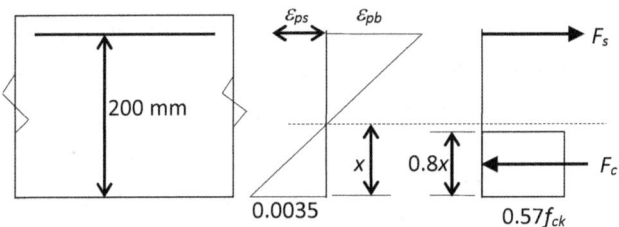

Assume tendons and bars yield:

Thus F_s= 0.87x0.9x300,000x28+0.87x500x201x9

= 6577 + 788 = 7365 kN.

F_s = F_c => 0.8x = 7365,000/(0.57x35x9000) = 41 mm,

=> x is 51 mm;

$\varepsilon_{pb} = 0.0035(d-x)/x = 0.0035(200-51)/51 = 0.010$

$\varepsilon_{ps} = 0.9\times0.56\times300,000/(165\times195\times10^3) = 0.005$

Total strain in strand = 0.005+0.010 = 0.015

Tendon yield strain is $0.87\times0.9\times300,000/(165\times195\times10^3)$

= 0.007 < 0.015. Hence tendon yields as assumed.

Now calculate M_R:

$M_R = F_s Z = F_s (d-0.4x) = 7,365,000(200 - 0.4(51))$

= 1322 kNm > 1005 kNm, OK

External column:

Due to M_2 moment in external column is substantially reduced. The <u>column</u> has to resist an ultimate moment of 416/2 = 208 kNm (as there is a column above and below; note that EC2 does not recommend any redistribution of *column* moments).

Note: If the <u>slab</u> were RC instead of PSC this <u>column</u> would have to be designed to resist an ultimate moment of (1.35(246+8) + 1.5x(196))/2 = 318 kNm.

However the slab should be checked to ensure that it is able to <u>transfer</u> the moment of 416 kNm.

50% redistribution will be used (see 3.7.4.2).

Moment to be transferred:

Moment at centre of support   = **- 416 kNm**

(before redistribution)

EC2, Fig 9.9 shows the breath of the effective moment transfer strip, b_e; all of the reinforcement should be concentrated into this width.

In this case, b_e = 600+450 = 1050 mm

The tendons are at mid height so their contribution will be small. Try 10H20s in the top of the slab,

T = 0.87x500x314x10 = 1,366 kN,

a = 0.8x = 1366x10^3/(1050x0.57x35) = 65 mm,

x = 65/0.8 = 82 mm; z = d - a/2 = 215-65/2 = 182 mm,

M_R = $T.Z$ = 1,366,000x182 = 248 kNm > 208 kNm => OK

Hence provide 10H20s at top of slab centered on this column (spacing about 100mm).

Slab: Middle Interior Span

Ultimate Sagging Capacity required = 524 kNm

Section 9m wide is reinforced with 28 strands

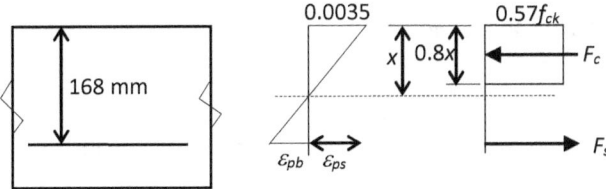

Assume tendons yield:

Thus $F_s = 0.87 \times 0.9 \times 300,000 \times 28 = 6,577$ kN.

$F_s = F_c \Rightarrow 0.8x = 6,577,000/(0.57 \times 35 \times 9000)$

$= 37$ mm, $\Rightarrow x$ is 46 mm $< d/2$ (= 90 mm)

$\varepsilon_{pb} = 0.0035(d-x)/x = 0.0035(168-46)/46 = 0.0093$

$\varepsilon_{ps} = 0.9 \times 0.56 \times 300,000/(165 \times 195 \times 10^3) = 0.0047$

Total strain in strand $= 0.0093 + 0.0047 = 0.014$

Tendon yield strain is $0.87 \times 0.9 \times 300,000/(165 \times 195 \times 10^3) = 0.007 < 0.014$. Hence assumption okay.

Calculate M_R:

$M_R = F_s Z = F_s (d - 0.4x) = 6,577,000(168 - 0.4(46)) = 984$ kNm

> 524 kNm $\Rightarrow$ OK

Notice from previous work, middle of <u>exterior</u> span will be okay too since $M_{SAG} = 873$ kNm

Internal Column

Because of M_2 internal columns are subject to somewhat <u>lower</u> moment.

$DL = 432 - 384 = -48$ kNm;

$M_2 = +47$ kNm; $LL2 = -122$ kNm

Ultimate moment $(1.35DL + 1.5LL + 0.9M_2) = -206$ kNm

If no M_2 then moment would be -248 kNm (= 1.35DL + 1.5LL).

Punching: typical internal column, (600x600)

Ultimate load: 1.35DL+1.5LL = 1.35(6.25+0.2) + 1.5(3+1+0.5) = 15.5 kN/m². Thus reaction is 9x9x15.5 = 1256 kN. Connection to be designed for β times this. Braced structure => V_t = 1.15x1256 Figure 8.1 of EC2. Thus V_t = 1444 kN.

Check at perimeter 1 (column perimeter): u_0

Ignoring tendons completely, d is the average of the H16 depths i.e., 250-25-16 = 209 mm

Design nominal stress, v_{Ed} = 1444,000/(600x4x209) = 2.9 N/mm² < $v_{Rd,max}$ (for cot θ = 45°) = 6.02 N/mm² => OK

Check at Perimeter 2: u_1 (2d from column face)

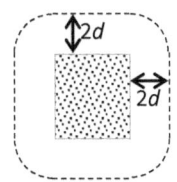

Make conservative assumption that tendons (which are near mid-height) are to be ignored completely.

V_{Ed} = 1444 kN, d = 209 mm (average of 2 layers).

v_{Ed} = 1444,000/((600x4+4x209xπ)209)

= 1444,000/((5026)209) = 1.37 N/mm^2.

(Note: we are allowed to subtract the loading inside the perimeter, but the amount is small so we'll ignore it.)

9H16s over 1350mm each way

=> 100A_{sl}/b_wd = 100x9x201/1350x209 = 0.64

$v_{Rd,c}$ = (0.18/1.5)(1+(200/d)$^{0.5}$)(f_{ck}100A_{sl}/b_wd)$^{1/3}$

$v_{Rd,c}$ = (0.12)(1.98)(22.4)$^{1/3}$ = 0.67 N/mm^2

Compare applied v_{Ed} = 1.37 N/mm^2

Thus $v_{Ed} > v_{Rd,c}$ hence provide shear reinforcement.

EC2 next requires us to determine position of outer control perimeter u_{out} at which v_{Ed} = $v_{Rd,c}$

Thus v_{Ed} = 1444,000/((600x4+ax209xπ)209) = 0.67 N/mm^2. Hence a = 12.1

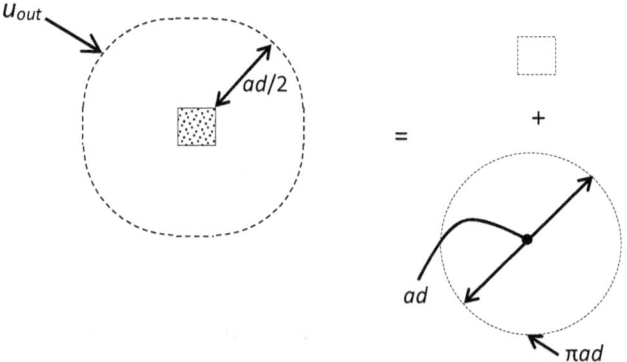

Hence u_{out} at 6d from column face (> 3d).

Outermost perimeter of links 1.5d inside this.

Shear reinforcement should be provided within area between column face and 1.5d inside outer control perimeter such that:

$A_{sw}=(v_{Ed}-0.75v_{Rd,c})s_ru_1/(1.5f_{ywd,ef})$ per perimeter

Where

s_r = radial spacing of links = 0.75d = 157 mm

$f_{ywd,ef}$ = effective strength of links = 250+0.25d = 250 + 52 = 302 N/mm^2

i.e. A_{sw} =(1.37-0.75x0.67)157x5026/(1.5x302)=1511 mm^2

H10 bars => 1511/78.5 = 19.2 say 20 bars per perimeter (minimum). Bar spacing limits will probably lead to the addition of links. Add H10 cage bars where necessary to anchor the links.

Perimeters at 0.5d, 1.25d, 2d, 2.75d, 3.5d, 4.25d, 5d from column face.

Additional Reinforcement

1. Place A7 BRC Mesh as anti-crack steel throughout bottom of slab between grids B and C (as tendons not near face of concrete).
2. Reinforcement parallel to walls (restraint reinforcement) A_s = 0.35% x h x $L/2$ (where L = wall length ≈ 9400 mm); = 0.35x250x9400/200 = 4113 mm^2 use 10H25 (5 top, 5 bot) (4910 mm^2)
3. Reinforcement between tendon anchorages for tendons in banded direction (distance ≈ 4.5 m)

4. Bursting reinforcement (designed as indicated in Chapter 8).

Deflection

No deflection is induced under DL because this load is balanced, and as slab is level no additional long-term deflection should theoretically be recorded due to DL.

To estimate elastic deflection due to LL we will use the formula for deflection of a flat plate (Naaman 1982):

$\Delta_l = kwl_a^4/E_c h^3 = 0.148*4.7*9^4/(34*10^6*0.25^3) = 0.009$ m

Allowable deflection $L/360 = 9/360 = 0.025$m => OK

Other Considerations:

Restraint: Maximum tendon extent < 50 m so no pour strips needed.

Construction (shrinkage): Consider pouring entire floor without construction joints.

Friction losses: All tendons are to be stressed from both ends. (20 m is usual limit for single-end stressing)

Progressive Collapse: Provide bottom bars at each column. Accident load = DL + 0.5LL $(0.25x25+0.2+0.5x4.5) = 8.7$ kN/m^2

Capacity of H16 = $A_s f_y$ = 201x500 = 100.5 kN.

No. of H16s required = 8.7x9x9/100.5 = 7.1 say 8.

Hence provide 8H16 bars. Bar length 4x35x16+400 = 2840 mm.

10.2. LONG-SPAN FLOOR TO EC2

An office building requires a long-span floor (16 m). The floor needs to have an exposed soffit, should be installed quickly, be durable, of low maintenance and should resist an imposed working load of 5 kN/m². Because of requirements of soffit finish, speed, durability and low maintenance it is decided to use precast concrete. Precast TT units prestressed with pre-tensioned steel will be used. *In-situ* structural concrete topping will be added after placing the unpropped precast unit. Design as a one-way uncracked member and limit tensile and compressive stresses to f_{ctm} and $0.45f_{ck}$ respectively under working load. At transfer use the above limits on the actual concrete strengths.

Solution

First, carry out a preliminary design to verify the main proportions. The approximate total depth of floor = $L/22$ = 16,000/22 = 727 mm. A suitable existing precast mould is available (shown below). This gives a total depth = 711 + 65 = 776 mm.

Loadings

Bare precast:

1)   Self-weight of precast unit = 7.87 kN/m.

2)   Structural Topping (100 mm at support and   65 mm at midspan, average 82.5 mm thick) = 4.95 kN/m.

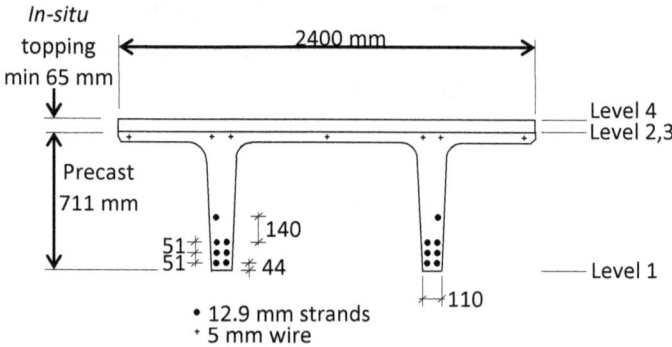

Composite Section:

3)   Ceiling and services at 1 kN/m^2 = 2.4 kN/m.

4)   Live Load = 5x2.4 = 12 kN/m.

Concrete:

1.   Precast Concrete f_{ck} = 40 N/mm^2 at transfer and lifting (when E_{cm} = 35 kN/mm^2 and f_{ctm} = 3.5 N/mm^2), and 50 N/mm^2 at 28 days (when E_{cm} = 37 kN/mm^2 and f_{ctm} = 4.1 N/mm^2). Unit delivered to site after 28 days.

2.   *In-situ* Concrete f_{ck} = 30 N/mm^2 at 28 days (when E_{cm} = 33 kN/mm^2).

Steel:

1.   Try 14 no. 12.9 mm nominal dia. strand (7 per rib) having characteristic breaking load of F_{pk} = 186 kN each and area of 100.7 mm^2 each.

2.   Try 7 no. 5 mm dia wires in the flange of F_{pk} = 30.3 kN each and area 19.6 mm^2 each.

Strand E_{steel} = 195 kN/mm^2; low relaxation strand 2.5% at 1000 hrs when stressed to 70% of breaking.

Properties of precast section (manufacturer's catalogue)

A_{pc} = 314,800 mm^2

y = 226 mm (centroid of precast measured from top of precast)

I_{pc} = 15.05x10^9 mm^4

Z_2 = 66.56x10^6 mm^3

Z_1 = 31.04x10^6 mm^3

Composite TT Construction

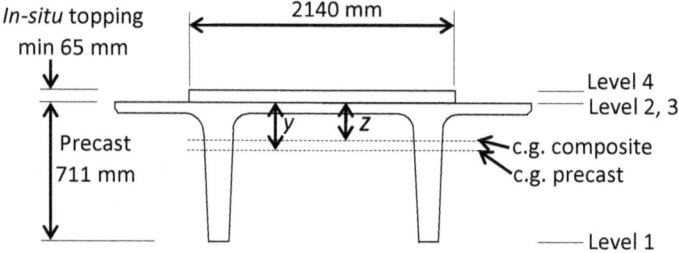

Transformed width of *in-situ* = 2400x33/37 = 2,140 mm.

Transformed area of *in-situ*, A_t = 2140x65 = 139,135 mm^2.

Properties of composite section

$A_{comp} = A_{pc} + A_t = 314,800 + 139,135 = 454,935$ mm^2

$z = (A_{pc}(y+65) + A_t x 65/2)/(A_{pc} + A_t)$

$$= (314,800 \times (226+65)+139,135 \times 65/2)/(454,935)=211 \text{ mm}$$

$$I_{comp} = I_{pc} + A_{pc}(y+65-z)^2 + I_{in-situ} + A_t(z-65/2)^2$$

$$= 15.05 \times 10^9 + 314,800(226+65-211)^2 + 2140(65)^3/12+139,135(211-65/2)^2 = 21.5 \times 10^9 \text{ mm}^4$$

$$Z_{2comp} = Z_{3comp} = I_{comp}/152 = 21.5 \times 10^9/146 = 147.3 \times 10^6 \text{ mm}^3$$

$$Z_{1comp} = I_{comp}/565 = 21.5 \times 10^9/565 = 38.0 \times 10^6 \text{ mm}^3;$$

$$Z_{4comp} = I_{comp}/211 = 21.5 \times 10^9/211 = 102 \times 10^6 \text{ mm}^3$$

Step 1: Check Ultimate Flexure

Load for ultimate condition:

$w = 1.35(7.87+4.95)+1.5(2.4+12) = 38.9 \text{ kN/m}$

Ultimate moment $M_{Ed} = wL^2/8 = 38.9 \times 16^2/8 = 1245 \text{ kNm}$

Ultimate resistance M_R :

Assume all tendons yield.

Ultimate tensile force = $14 \times 0.87 \times 0.9 \times 186 = 2,039 \text{ kN}$

CG of tensile force = 122 mm from bottom face.

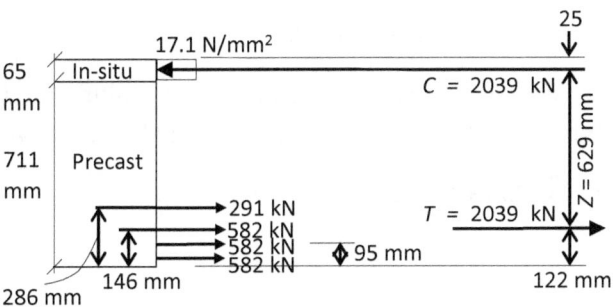

Compression stress in *in-situ* = 0.57x30 = 17.1 N/mm^2

Force in *in-situ* = 2039 kN

Depth of compression zone in *in-situ* = 2,037,000/(2400x17.1) = 50 mm

Centroid of compressive force = 25 mm from top

M_R = *T.Z* = 2,039x0.629 = 1,282 kNm > 1,245 kNm   OK

Check <u>uppermost</u> strands to see if they yield:

Each strand stressed to 70% ultimate and there are say 30% losses, so ε_{ps} = 0.9x0.7x0.7x186/(100x195) = 0.0041

Additional strain due to bending = 427x0.0035/63 = 0.023

Hence total strain = 0.0041+0.023 = 0.027

Yield strain = 0.9x0.87x186/(100x195) = 0.0075

Hence assumption that all strands yield is correct.

Step 2: Transfer

Properties for transfer check:

Each strand and wire stressed to 70% ultimate, i.e., 0.7x186 = 130.2 kN and 0.7x30.3 = 21.2 kN respectively.

P_j = 14x130.2+7x21.2 = 1,971 kN

Centroid of P_j is 164 mm from <u>bottom</u> of precast (see below), i.e., eccentricity of prestressing force, e = 711-226-164 = 321 mm below the centroid of the precast.

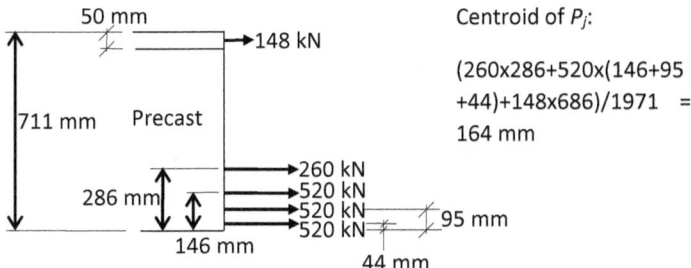

Centroid of P_j:

(260x286+520x(146+95 +44)+148x686)/1971 = 164 mm

Step 3: Two-Point Lifting from mould

At lifting f_{ck} = 40 N/mm^2.

Unit lifted at 1/5 points (i.e. 3.2 m from ends).

Allow 50% impact allowance => M_{top} = 1.5x7.87x3.2^2/2 = 60.5 kNm

Top reinforcement 7 wires => Tension at yield, T = 7x0.87x0.9x30.3 = 166 kN.

Effective depth d = 711-25 = 686 mm

$T = C$ => 166,000 = 0.57x40x102x2xs => s = 36 mm,

Lever arm, z = 0.95d = 652 mm

$M_R = Tz$ = 166x0.652 = 108 kNm > 60.5 kNm

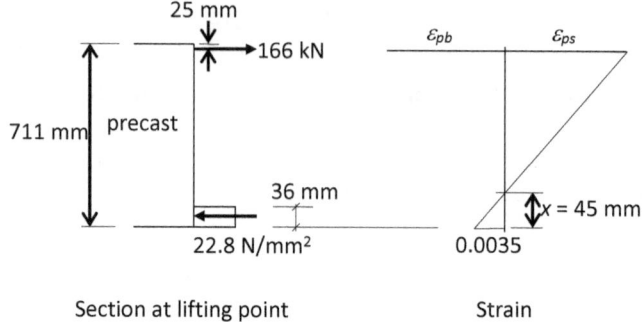

Section at lifting point   Strain

Calculate losses

We'll use the procedure suggested in Abeles (1981) to estimate the losses.

(1)   Elastic Shortening:

$$\Delta P_{ES} = n_T f_{cp} A_{ps}$$

where n_T is modular ratio at transfer and f_{cp} is stress in concrete at level of steel.

$n_T = E_{steel}/E_{conc} = 195/35 = 5.6$

Estimating prestressing force to be $0.9P_j$ where $P_j = 14 \times 0.7 \times 186 + 7 \times 0.7 \times 30.3 = 1,971$ kN

f_{cp} = Axial Effect + Bending Effect + Dead-weight

$= 0.9P_j / A_{pc} + 0.9P_j\, e^2/ I_{pc} - M_o\, e/ I_{pc}$ $=$
$0.9 \times 1,971 \times 10^3 (1/314,800$ $+$ $321^2/15.05 \times 10^9)$ $-$
$(7.87 \times 16^2 \times 10^6/8) \times 321/15.05 \times 10^9$

$= 1,774 \times 10^3 (1/314,800 + 1/145,961) - 5.38 = 17.7 - 5.38 = 12.4$ N/mm^2

Thus $\Delta P_{ES} = n_T f_{cp} A_{ps} = 5.6 \times 12.4 \times 1547 = 107$ kN

(2)   Relaxation at 1000-hr = 2.5% => long-term value
= 1.2×2.5 = 3% of 1,971 kN = 59 kN

(3)   Shrinkage Loss

$$\Delta P_{SH} = \varepsilon_{SH} E_{ps} A_{ps} = 400 \times 10^{-6} \times 195 \times 1547$$

$$= 121 \text{ kN}$$

(4)   Creep Loss

$$\Delta P_{CR} = \Phi \, n_L f_{cp} A_{ps}$$

where f_{cp} is the concrete stress at the level of the tendons due to all sustained loads. $n_L = E_{steel}/E_{conc} = 195/37 = 5.3$. Assume Φ is 2.

$f_{cp} = 0.8 P_j / A + 0.8 P_j e^2 / I_{pc} - M_{DL+ADL} \, e / I_{pc}$

$= 0.8 \times 1,971 \times 10^3 (1/314,800 + 321^2/15.05 \times 10^9) -$

$\qquad\qquad (12.8 \times 16^2 \times 10^6 / 8) 321 / 15.05 \times 10^9$

$= 1,577 \times 10^3 (1/314,800 + 1/146,058) - 8.7$

$= 5.0 + 10.8 - 8.7 = 7.1 \text{ N/mm}^2$

Hence $\Delta P_{CR} = 2 \times 5.3 \times 7.1 \times 1,547 = 116$ kN

Thus effective prestress:

$P_e = 1,971 - 107 - 59 - 121 - 116 = \mathbf{1,568 \ kN}$

Initial prestress:

$P_i = 1,971 - 107 = \mathbf{1,864 \ kN}$

Check:

- Elastic: 107 kN of 1,971 kN => 5.4 %

- Relaxation: 49 kN of 1,971 kN => 2.5%

- Shrinkage: 121 kN of 1,971 kN => 6.1%

- Creep: 116 kN of 1,971 kN => 5.8%

- Total = 19.8%

Comment: Small creep loss, but not unusual. Recall: Losses have little effect on ultimate capacity.

Transfer stresses:

Stress at <u>bottom</u> of section at midspan (allow counteraction at midspan):

$f_1 = P_i/A + P_i e/Z_1 - M_0/Z_1$

$= 1,864x10^3/314,800 + 1,864x10^3x321/31.04x10^6$
$\qquad - 7.87x16^2x10^6/(31.04x10^6x8)$

$= + 5.92 + 19.2 - 8.11 = + 17.0 \ N/mm^2$

Allowable compression $0.45f_{ck} = 0.45x40 = 18 \ N/mm^2$

Hence OK

Check transfer stresses at <u>end</u>:

Stress at <u>top</u> of section:

$f_2 = P_i/A - P_i e/Z_2$

$= 1,864x10^3/314,800 - 1,864x10^3x321/66.56x10^6$

= + 5.92 − 8.99 = - 3.07 N/mm^2 (tension).

Allowable tension f_{ctm} = 3.5 N/mm^2,

Hence ok

Check transfer stresses at <u>bottom</u> at end:

$f_1 = P_i / A + P_i e / Z_2$

= 1,864x10^3/314,800 + 1,864x10^{3}x321/31.04x10^6

= + 5.92 + 19.3 = + 25.2 N/mm^2 .

Allowable compression 0.45f_{ck} = 18 N/mm^2.

Hence <u>de-bond</u> some tendons close to ends.

Try de-bonding the four lowest strands:

P_i becomes 1,864-521 = **1,343** kN.

After de-bonding:

Ignoring effect on eccentricity e (e will be reduced):

$f_1 = P_i / A + P_i e / Z_2$

= 1,343x10^3/314,800 + 1,343x10^{3}x321/31.04x10^6

= + 4.27 + 13.9 = + 18.2 N/mm^2 (compression).

Allowable 0.45f_{ck} = 18 N/mm^2. This is close enough.

De-bond for 5 m length. M_o at this point will be about 80% midspan value.

Stage 1:

Precast unit placed on neoprene bearings and *in-situ* topping poured.

Calculate stresses at midspan:

Bottom of precast (level 1):

$f_1 = P_e/A + P_ee/Z_1 - M_{DL+ADL}/Z_1$

$= 1,568 \times 10^3/314,800 - 1,568 \times 10^3 \times 321/31.04 \times 10^6 + (7.87+4.95) \times 16^2 \times 10^6/8 \times 31.04 \times 10^6$

$= + 4.98 + 16.2 - 13.2 = + 7.98$ N/mm^2 (compression)

$< 0.45f_{ck} = 22.5$ N/mm^2

Top of precast (level 2):

$f_2 = P_e/A - P_ee/Z_2 + M_{DL+ADL}/Z_2$

$= 1,568 \times 10^3/314,800 - 1,568 \times 10^3 \times 321/66.56 \times 10^6 +$

$(7.87+4.95) \times 16^2 \times 10^6/8 \times 66.56 \times 10^6$

$= + 4.98 - 7.56 + 6.16 = + 3.58$ N/mm^2

Stage 2:

Topping hardened. Additional load applied (ceiling, live load).

Bottom of precast (level 1):

$f_1 = -M_{ADL+LL}/Z_{1comp}$

$= - (2.4+12) \times 16^2 \times 10^6/(8 \times 38 \times 10^6) = - 12.1$ N/mm^2 (tension)

Hence total $f_1 = + 7.98 - 12.1 = - 4.12$ N/mm^2 (tension)

Allowable tensile stress f_{ctm} = 4.1 N/mm^2. Hence OK.

$f_2 = M_{ADL+LL} / Z_{2comp}$

= (2.4+12)x16^{2}x10^6/8x147.3x10^6

= 3.13 N/mm^2 (compression)

Hence total f_2 = 3.58 + 3.13 = 6.71 N/mm^2 (compression)

Stresses in *in-situ* topping:

$f_3 = +n\ M_{ADL+LL} / Z_{3comp}$

= +(33/37)x(2.4+12)x16^{2}x10^6/(8x147.3x10^6)

= + 2.80 N/mm^2 (compression)

$f_4 = +n\ M_{ADL+LL} / Z_{4comp}$

= +(33/37)x(2.4+12)x16^{2}x10^6/(8x102x10^6)

= + 4.03 N/mm^2 (compression)

Midspan Stresses under full working load

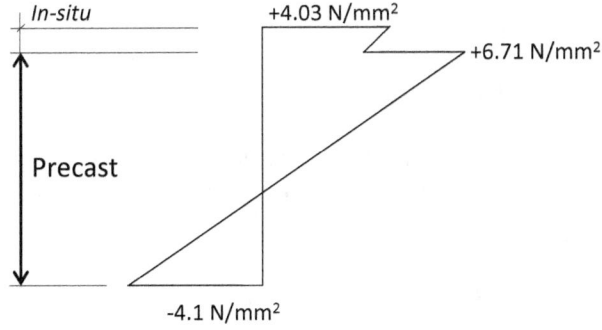

Step 4: Check Interface Shear

Check horizontal shear on interface between topping and precast beam:

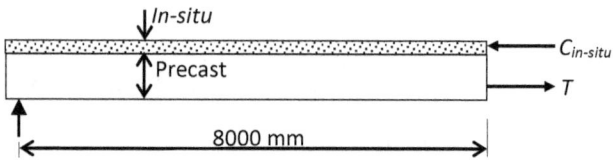

Value of C in *in-situ* at ultimate = 0.57x30x2400x50 = 2,052 kN.

Hence average horizontal shear stress = 2,052,000/(2400x8000) = 0.107 N/mm^2.

Load is *UDL*, so beam shear is triangular; thus interface shear assumed as to be of triangular distribution too. Hence maximum interface shear stress = 2x0.107 = 0.214 N/mm^2

From 6.2.5 of EC2

$v_{Rd,i} = 0.4f_{ctd}$ where $f_{ctd} = 2/1.5 = 1.33$ for $f_{ck} = 30$ N/mm^2 so $v_{Rd,i} = 0.53$ N/mm^2, thus ok

Step 5: Check "Beam" Shear

The worst value of shear force occurs at the end of beam. It is reasonable and conservative to ignore prestress entirely and design as an *RC* member.

Effective depth = 629 + 25 = 654 mm

Ultimate *UDL* w = 38.4 kN/m

So design value of V i.e. V_{Ed} is taken at d from the face = (8-0.15-0.654)x38.4 = 276 kN.

v_{Ed} = 276,300/(2x110x654) = 1.92 N/mm^2.

$v_{Ed,z}$ = 1.92/0.9 = 2.13 N/mm^2

Max allowable stress $v_{Rd,max}$ is 5.52 N/mm^2 at θ = 21.8°. Hence θ = 21.8°.

$A_{sw}/s = v_{Ed,z}b_w/f_{ywd}\cot\theta$

Provide 2 legs H8 links per rib (i.e., 4 legs per TT), => spacing required s = 0.87x500x200x2.5/(2.13x110) = 928 mm, say 425 mm centres (< 0.75d = 445). Provide this reinforcement over the full span.

4no. H10 cage bars are needed to anchor these links.

(Check links are at least nominal: f_{ck} = 40 N/mm^2; => $\rho_{w,min}$ = 1.13x10^{-3}; A_{sw}/sb_w = 0.00113 => s = 200/(2x110x0.00113) = 804 mm > 425 mm. Thus provide H8 links at 425 mm)

Shear at reduced section

The section will be dapped at the end. A pocket of 225 mm deep is proposed. Check shear at reduced section.

Effective depth, d = 402 mm (see below).

Take width of rib as 110 mm.

Thus v_{Ed} = (38.4x16/4)/(402x110) = 3.47 N/mm^2.

Limit is $v_{Rd,max}$ for $\cot\theta = 1$ = 8 N/mm^2 => ok

Model for pocket less than depth/3:

Concrete struts C_1 and C_2 form if ties T_1, T_2 and T_3 present. Pocket 225 mm x 150 mm.

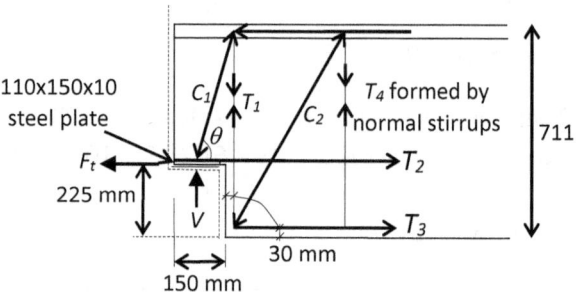

Truss action at support

$V = 38.9 \times 10^3 \times 16/4 = 156$ kN; Friction $F_t = 0.4V = 62.4$ kN

EC2 assumes strut C_2 is at 45°

Check bearing stress:

Nominal bearing 130 mm (allow 20 mm gap to *in-situ*).

Net bearing length (i.e. parallel to span) = 130 − 15 (spalling of precast) − 25 (spalling of *in-situ*) − 15 (tolerance of precast length) = 75 mm.

Net bearing width = 110 mm.

Allowable stress = $0.567 f_{ck}$ = 28.3 N/mm².

Bearing stress = $156,000/(110 \times 75)$ = 18.9 N/mm^2 < 28.3 N/mm^2 => ok

Consider force triangle formed by C_1, T_1 and T_2. Make reasonable guesses as to bar diameters.

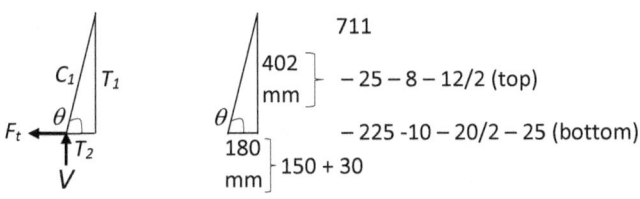

711

402 mm $\Big\} - 25 - 8 - 12/2$ (top)

$- 225 - 10 - 20/2 - 25$ (bottom)

180 mm $\Big\} 150 + 30$

Tan $\theta = 402/180 = 2.33$ => $\theta = 65.9$ degrees

Strut C_1:

$C_1 sin\theta = T_1 = V => C_1 = 155/sin65.9° = 170$ kN.

Allowable stress at ultimate = $0.6(1-f_{ck}/250)f_{ck}/1.5 = 16$ N/mm^2

Thus width, $w = 170,000/110 \times 16 = 97$ mm.

EC2 limits stress block depth of recessed section to $0.6d_r$ = $0.6 \times (402) = 241$ mm

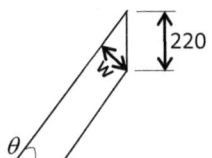

220

thus w is limited to $241cos\theta = 98$ mm > 97 mm => ok

Tie T_1

$T_1 = V \Rightarrow T_1 = 155$ kN $\Rightarrow A_s = 155,000/(0.87 \times 500)$

$= 356$ mm^2; Provide 3H10 (2 leg) links ($= 471$ mm^2)

at 50 mm crs starting at 25 mm from pocket.

Tie T_2

$T_2 = C_1 \cos\theta = 170 \cos 65.9° = 69.4$ kN $\Rightarrow$ tie force $= T_2 + F_t$
$= 69.4 + 62.4 = 132$ kN

Use high yield bars but with stress limits appropriate to mild steel (as the bars are to be welded to a mild steel plate) $\Rightarrow$ area required $= 132,000/(0.87 \times 250) = 607$ mm^2

Use 2H20 ($= 628$ mm^2) welded to end plate in order to achieve full anchorage.

Strut C_2 and Tie T_3

The diagonal compressive force C_2 is given by $V/\cos 45° = 156,000/\cos 45° = 221$ kN.

We check the width of the strut in the same way as C_1. This time, assuming T_3 consists of H16 bars, $d = 711-35-16-16/2 = 652$ mm. Thus $0.6d = 391$ mm.

Hence allowable width, $w = 391 \sin 45° = 277$ mm.

Actual w required $= 221,000/110 \times 16 = 125$ mm $\Rightarrow$ ok

Now $T_3 = V$ hence area required $= 356$ mm^2

Provide 2H12 bars + 1H16 (427 mm^2)

Nominal bars

Provide nominal horizontal and vertical bars on the face of the element in reduced depth area: $A_{req'd}$ = 0.2%

(711-225)x110x0.002 = 107 mm^2

Provide 2H10 U-bars Horiz & Vert (latter welded to plate) (157 mm^2)

Check that at least the area of vertical stirrups suggested by ACI-318 is provided ignoring the pocket, i.e. A_t = $0.021 P_i h / f_s l_t$ spread over end h/5. Thus A_t = 0.021x(1864000/2)x711/(140x903) = 110 mm^2 per rib spread over end 711/5 = 142 mm. Now 2 legs of 10mm dia bar is 157 mm^2 > 110 mm^2 => we have provided more than enough.

Summary

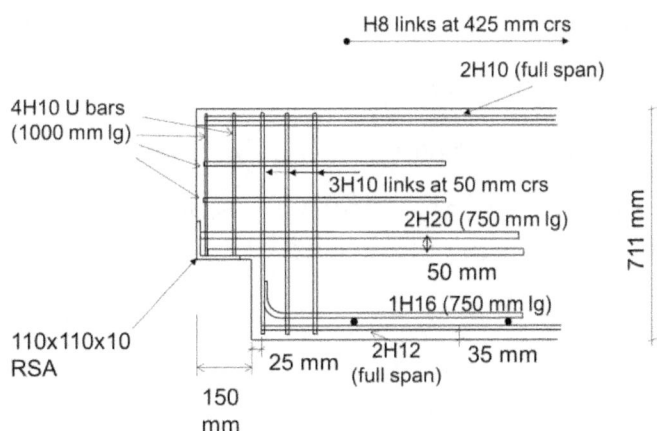

Step 6: Check Deflection

It can be easily shown that under quasi-permanent loads f_1 is within the allowable stresses for uncracked structures, so the gross I will be used to estimate deflection.

Immediate midspan upward deflection due to prestress alone:

$\delta_p = P_i e_{ps} L^2/(8E_c I_{pc}) = 1343 \times 10^3 \times 321 \times (16,000)^2/$

$8 \times 35 \times 10^3 \times 15.05 \times 10^9 = 26.2$ mm

Once the beam is prestressed the self-weight acts:

$\delta_0 = 5 w_0 L^4/(384 E_c I_{pc})$

$= 5 \times 7.87 \times (16,000)^4/384 \times 35 \times 10^3 \times 15.05 \times 10^9 = 12.7$ mm

Hence immediate deflection is 26.2-12.7 = 13.5 mm (upward) < L/250 = 16,000/250 = 64 mm, ok

Incremental deflection due to LL:

$= \delta_0 = 5 w_0 L^4/(384 E_c I_{comp})$

$= 5 \times (12+2.4) \times (16,000)^4/384 \times 37 \times 10^3 \times 21.5 \times 10^9$

$= 15.4$ mm $= L/1035 < L/250$, Hence okay

CHAPTER 11: FURTHER EXAMPLES

Q1: *(a) What are the stresses due to prestress alone in this cranked slab?*

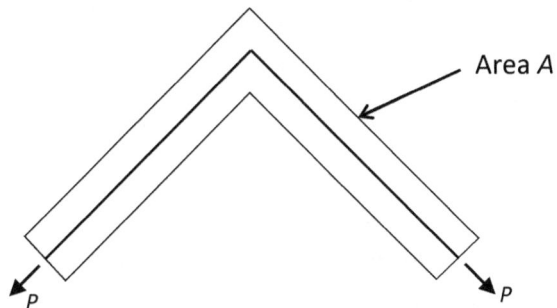

Area *A*

P

P

Note: Straight concentric tendon

Q1: *(b) What are the stresses due to prestress alone at the top of this semi-circular arch?*

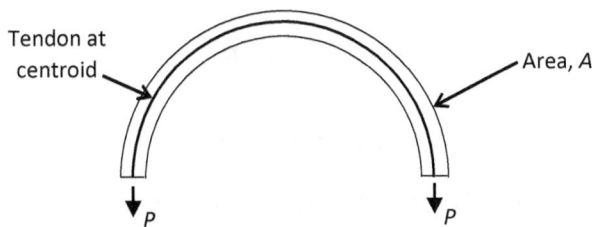

Tendon at centroid

Area, *A*

P

P

Q1 (a) Answer

Free body of **concrete**:

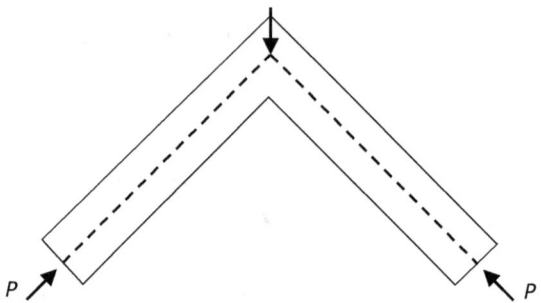

Concentrated force at kink point requires reactions to maintain equilibrium. These forces are *P*.

Imagine a perpendicular cut through the right leg:

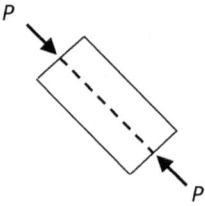

Free body shows we have *P/A* at each section.

General Rule: If tendon follows the centroid then there is no bending moment induced (e.g. brake cable on a bicycle).

Q1 (b) Answer

Net effect: Stresses throughout P/A. We can see this by applying our rule from (a). (Imagine any curved surface as consisting of many short straight portions with sharp kinks as in (a)). The tendon follows the centroid so stresses throughout are P/A.

Q2: *In prestressed concrete structures we must calculate stresses in service. Why do we not worry about service stresses in conventional RC?*

Q2: Answer

1. We expect permanent cracks (RC **must** crack in order to work). Steel in conventional RC carries virtually no force until the concrete around it cracks. Codes require us to check crack widths in some instances.
2. There is lots of steel in RC compared to PSC. This steel is usually relatively closely spaced and close to the surface. Thus it is effective at controlling the width of these cracks.

Q3: *Calculate the stresses at midspan due to prestressing and due to a UDL of 15 kN/m (including self-weight of beam).*

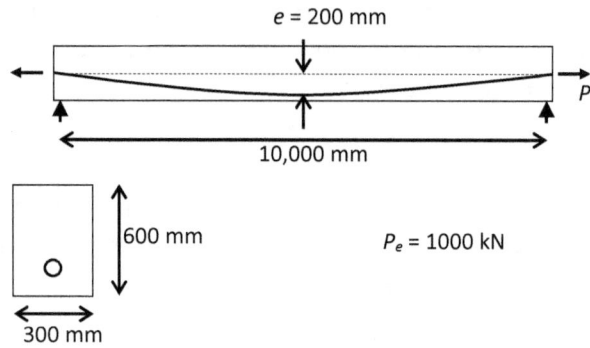

e = 200 mm

10,000 mm

600 mm

300 mm

P_e = 1000 kN

Q3: Answer

Due to prestressing:

$P/A = 1000 \times 10^3/(300 \times 600) = 5.5$ N/mm^2

$Pe/Z = 1000 \times 10^3 \times 200 \times 6/(600^2 \times 300) = 11.1$ N/mm^2

Stress due to UDL:

$M/Z = 15 \times 10000^2 \times 6/(8 \times 600^2 \times 300) = 10.4$ N/mm^2

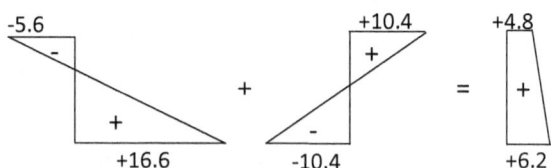

-5.6

+16.6

+10.4

-10.4

+4.8

+6.2

Q4: *Would it be advantageous to raise the anchors at the simply supported end of a prestressed beam to increase the drape and so the equivalent load further?*

Q4: Answer

No. It can be shown that the additional negative span moment due to increased sag is **exactly** offset by an opposing positive span moment introduced by anchor eccentricity. Thus there is no benefit in raising the anchors above the centroid.

Illustration:

(a) Find midspan bottom stress when $\delta = 400$ mm

(b)

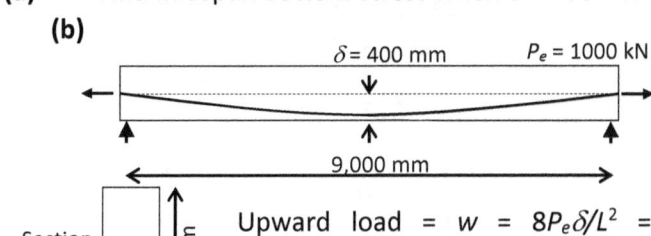

$\delta = 400$ mm $P_e = 1000$ kN

9,000 mm

Section at end

900 mm

300 mm

Upward load $= w = 8P_e\delta/L^2 = 8*1,000*0.4/9^2 = 39.5$ kN/m

Midspan moment $= 39.5*9^2/8 = 400$ kNm => Bottom stress $= f_1 = P_e/A + M/Z = 1,000*10^3/(300*900) + 6*400*10^6/300*900^2 = 3.7 + 9.9 = 13.6$ N/mm^2

(c) Now suppose the tendon is raised so that the drape is increased to 600 mm.

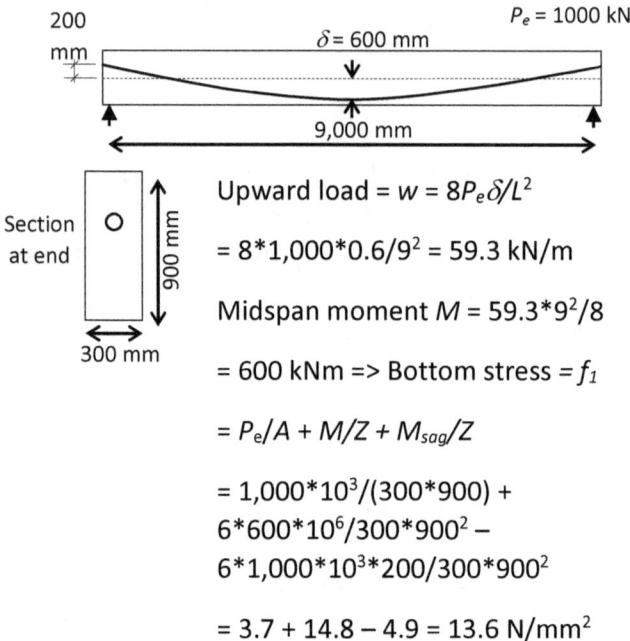

Upward load = $w = 8P_e\delta/L^2$

= 8*1,000*0.6/9² = 59.3 kN/m

Midspan moment M = 59.3*9²/8

= 600 kNm => Bottom stress = f_1

= $P_e/A + M/Z + M_{sag}/Z$

= 1,000*10³/(300*900) + 6*600*10⁶/300*900² − 6*1,000*10³*200/300*900²

= 3.7 + 14.8 − 4.9 = 13.6 N/mm²

Apart from there being no economic benefit in raising the ends to increase the drape, doing so is likely to make the detailing of the anchorage more difficult.

Q5: *The following two prestressed beams are identical except for the tendon profiles. Are the stresses at midspan due to prestress alone different?*

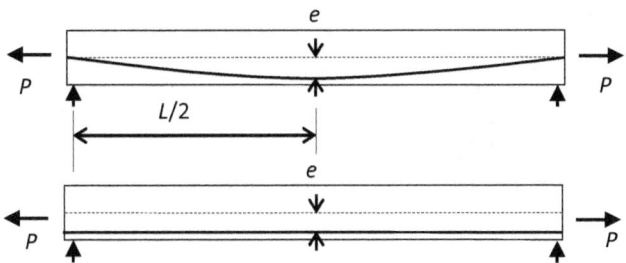

Q5: Answer

The stresses are the same.

Beam 2: $P/A + Pe/Z$

Beam 1: $w_{upward} = 8P\delta/L^2 = 8Pe/L^2$; $M_{max} = w_{upward} L^2/8 = Pe$

so stress is $P/A + Pe/Z$

Thus stresses are unaffected by the profile of tendon elsewhere. Only the conditions *at the section* matter.

Q6: *Calculate the primary and secondary moments due to prestress.*

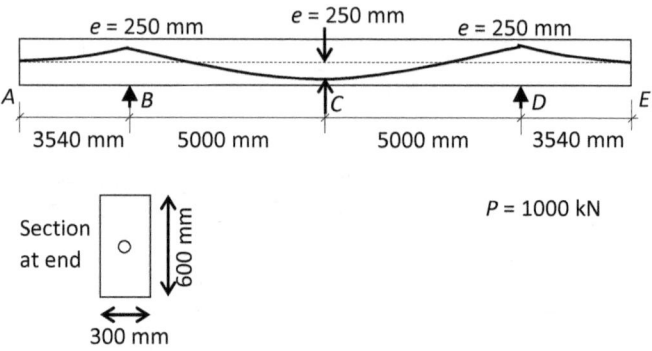

Q6: Answer

Primary moment, $M_1 = P.\, e$, where e is eccentricity at the point.

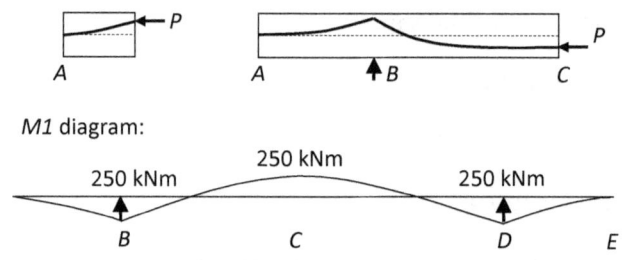

M1 diagram:

A   250 kNm   250 kNm   250 kNm

A   B   C   D   E

Calculate the total moment due to prestressing, M_{TOT}:

Cantilever span: Equivalent load $w_{can} = 2P\delta_{can}/L_{can}^2 = 2*1000*0.25/3.54^2 = 40$ kN/m

Main span: Equivalent load w_{main} = $8P\delta_{main}/L_{main}^2$ = $8*1000*0.5/10^2$ = 40 kN/m (upwards).

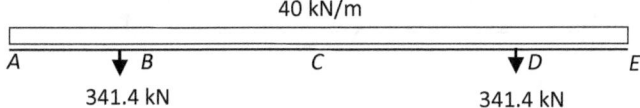

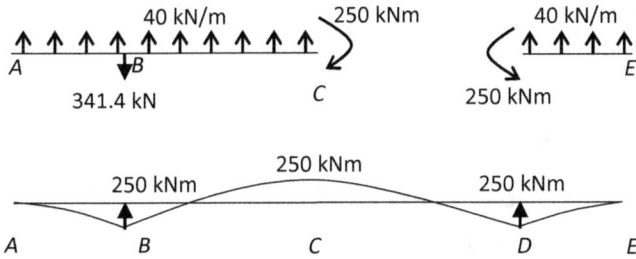

Thus $M_1 = M_{TOT} => M_2 = 0$

Actually, we could have saved ourselves some time; we could have observed that this structure is statically determinate and so $M_2 = 0$.

Q7: *Calculate the balanced load in the following case. Calculate the stresses at midspan when a UDL of 16 kN/m (including self-weight of beam) is applied.*

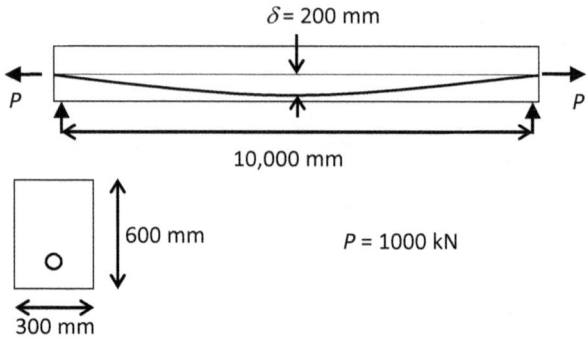

Q7: Answer

Balanced load $w = 8P\delta/L^2 = 8\times1000\times0.2/10^2 = 16$ kN/m, i.e., UDL = 16 kN/m

Stress due to prestressing and UDL of 16 kN/m:

The beam is under no bending stress, just axial compression due to P. Thus stresses everywhere:

$P/A = 1000\times10^3/300\times600 = 5.55$ N/mm^2

+   +5.55 N/mm^2

Q8: *On the following beam what cable profile would you suggest to best counteract the loads shown?*

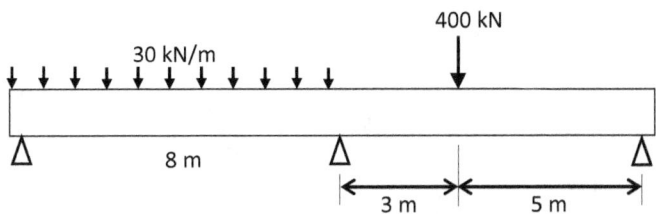

Q8: Answer

The UDL due to dead weight is 0.3x0.6x24 = 4.3 kN/m so is much smaller than the imposed loads and so can be ignored when we are deciding the profile. A UDL is best countered by a parabolic tendon and a point load by a harped tendon. The eccentricities are as close to the surface as cover limits allow and the tendon should be roughly in accordance with the ultimate BMD.

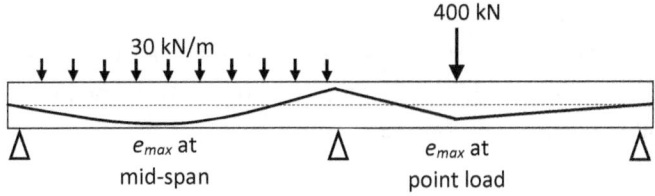

Q9: *Why is post-tensioned concrete particularly suitable for transfer beams?*

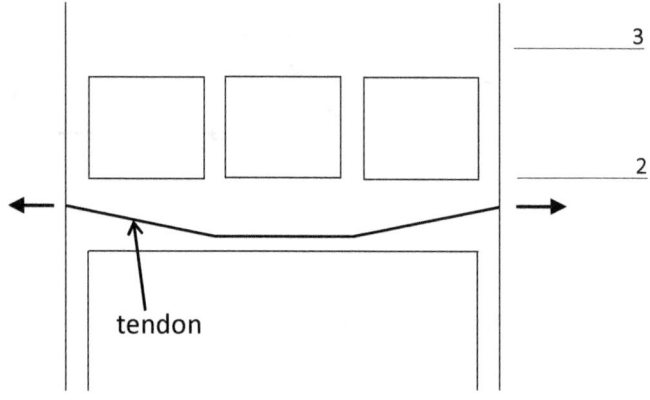

tendon

Q9: Answer

Much of the load which is applied to transfer structures is dead load, so this load can be balanced. Thus the beam will experience no bending from this load.

In addition, the upper structure is vulnerable to relative vertical movement of its supports; post-tensioning allows deflections to be substantially reduced.

Q10: *A simply supported member prestressed with a parabolic tendon is shown below. Calculate 1) the upward loading due to prestress, and 2) the load to be used for the ultimate design. DL = 6.7 kN/m, LL = 10 kN/m.*

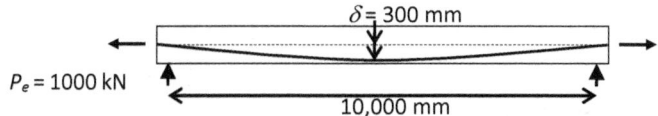

$\delta = 300$ mm

$P_e = 1000$ kN

10,000 mm

Q10: Answer

1. Upward load $w = 8P_e\delta/L^2 = 8\times1000\times0.3/10^2$

    = **24 kN/m**.

2. Ultimate load w_{ult} = 1.35DL+1.5LL = 1.35x6.7+1.5x10 = 24 kN/m. Hence design beam to resist a downward load of **24 kN/m**.

Clearly the upward load from the tendons <u>cannot</u> be included at ultimate. If included, then it leads to absurd conclusions. Here the conclusion would be that there is no strength requirement and so a beam of arbitrary strength would suffice. This is obviously wrong. Instead we should only include effects that change reactions. In this case here are no such effects as the beam is statically determinate. Hence the ultimate strength check is to check against the moment and shear due to 24 kN/m.

Q11: *Usually, the prestressing contractor submits tendon extensions to the project's structural consultant. How close to the theoretical extensions should the actual ones be? Should the tendon be re-stressed if the discrepancy is too much?*

Q11: Answer

The UK organization Cares has produced a specification for post-tensioned construction (see Cares 2011). It recommends that extensions are within +/- 15% for any individual tendon and within +/- 6% for any group of tendons.

Avoid re-stressing unless absolutely necessary. This is because when the tendon is anchored, wedges bite into tendon reducing its cross section, so there is a risk of breaking the tendon if it is re-stressed--thus it is dangerous. Also it should be borne in mind that as prestressing steel is high carbon steel, complete de-stressing and re-stressing is likely to reduce tendon's strength.

Q12: *A flat slab is constructed with a tendon profile as follows:*

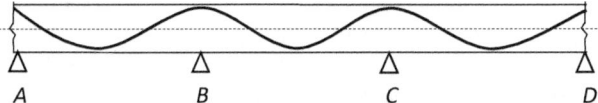

A late change results in column B being deleted. Clearly columns A and C need to be re-designed, but what about the slab?

Q12: Answer

The constructed configuration is this:

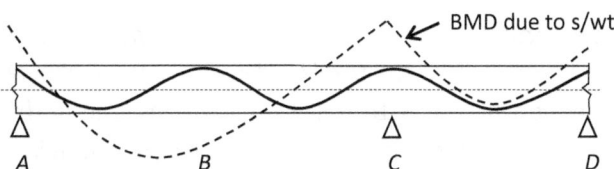

The previously calculated stresses would obviously be incorrect but more importantly there is little ultimate moment capacity at section B as the tendon is high in the section when it should be low in the section. Collapse is virtually certain. In fact this example is based on L'ambiance plaza (see for example MacAlevey 2010).

Q13:

Consider a 250 mm thick slab supporting the following:

1. Roof where LL = 0.75 kN/m²or

2. Warehouse where LL = 20 kN/m²

There is no ADL.

Use this example to explain why, in principle, PT is more efficient for low LL structures.

Q13: Answer

Permanent load = DL = 0.25x25 = 6.25 kN/m² in both cases.

1. Roof: Total service load = 6.25 + 0.75 = 7.0 kN/m²

The percentage of total load that is permanent = 6.25/7 = 89%. Thus most of the total load can be balanced by PT. The section only has to resist 11% of the loading (the unbalanced load if PT balances all of the permanent load).

2. Warehouse:

Total service load = 26.25 kN/m²

Percentage of total load that is permanent = 6.25/26.25 = 24%. Thus if the PT balances all of the permanent load the section has to resist about 78% of the loading (the unbalanced load). Hence a significant portion of the LL must be balanced too in order for the final stresses under total load to be acceptable. Thus more PT and/or deeper members must be used.

Q14 (Gilbert 1990): *The continuous beam below is 400 mm wide and 900 mm deep. The prestressing force of 1800 kN is assumed constant over the length of the beam. Find the primary and secondary moments.*

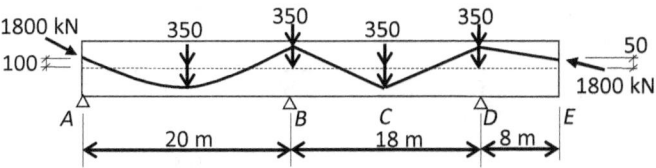

Q14: Answer

Equivalent Loads:

Span AB: $w = 8P\delta/L^2 = 8\times1800\times(0.25/2+0.45)/20^2$

$= 20.7$ kN/m

Span BD: $W = 4P\delta/L = 4\times1800\times(0.35+0.35)/18 = 280$ kN

Span DE: $W = P\theta = 1800\times(0.3/8) = 67.5$ kN

The continuous beam is then analysed under these loads as if they were external loads.

Moment distribution is used here. The loads on the concrete and analysis is as follows: (The loads due to reverse curvature at the supports are not shown as they play no part in the bending of the beam).

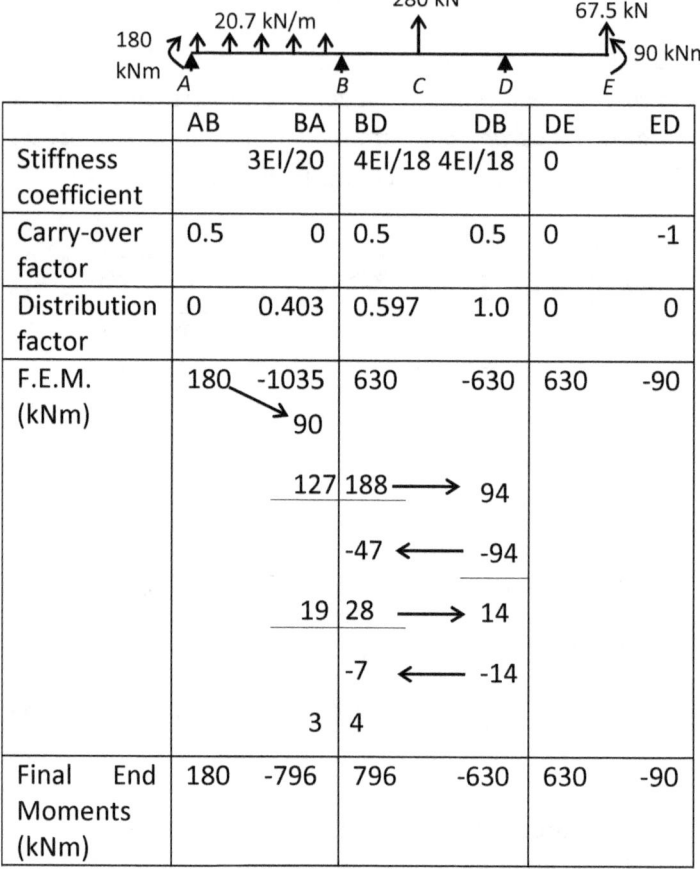

	AB	BA	BD	DB	DE	ED
Stiffness coefficient		3EI/20	4EI/18	4EI/18	0	
Carry-over factor	0.5	0	0.5	0.5	0	-1
Distribution factor	0	0.403	0.597	1.0	0	0
F.E.M. (kNm)	180	-1035 ↘ 90	630	-630	630	-90
		127	188 →	94		
			-47 ←	-94		
		19	28 →	14		
			-7 ←	-14		
		3	4			
Final End Moments (kNm)	180	-796	796	-630	630	-90

Total moment due to prestress, M_{TOT} (kNm):

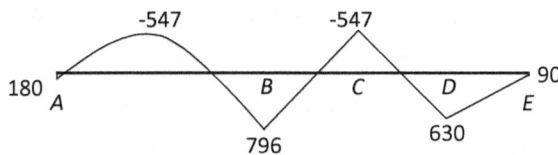

e.g. Moment at middle of Span AB = $-(20.7 \times 20^2/8)$ + $(796+180)/2 = -547$

The primary moment $M_1 = Pe$ is as follows:

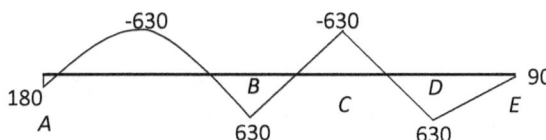

e.g. $M_1 = Pe = 1800 \times 0.35 = 630$ kNm

The secondary moment diagram $M_2 = M_{TOT} - M_1$ (kNm) is as follows:

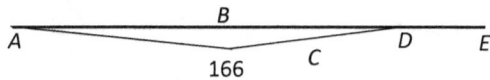

e.g. $M_{TOT} - M_1 = 796 - 630 = 166$ kNm

REFERENCES

Abeles, P.W., Bardhan-Roy, B.K. *Prestressed Concrete Designer's Handbook*, 3rd Edition, Viewpoint, London, 1981.

American Concrete Institute, ACI-318-14, *Building Code Requirements for Structural Concrete*, 2014.

Bennett, D.F.H., Billingham, A., "Exchange Tower (Project Profile)", British Cement Association, 1991.

Bobrowski, J., Bardhan-Roy, B.K., "A method for calculating the ultimate strength of reinforced and prestressed concrete beams in combined flexure and shear", *The Structural Engineer*, 1969, Vol. 47, No. 5, May, p197-209.

Bobrowski, J., *Origins of safety in concrete structures*, University of Surrey, 1982, PhD thesis.

Billington, D. P., "Historical Perspectives on Prestressed Concrete", *PCI Journal*, 2004, January-February, p14-30.

British Standards Institution BS EN 1992-1-1:2004, *Eurocode 2: Design of concrete structures — Part 1-1: General rules and rules for buildings*, 2005.

Cares, "Model specification for bonded and unbonded post-tensioned concrete floors", 3rd Edition, UK, 2011.

Elliott, K.S., Jolly, C., *Multi-storey Precast Concrete Framed Structures*, Wiley Blackwell Science, 2nd Edition, London, 2013.

Gilbert, R.I., Mickleborough, N.C., *Design of Prestressed Concrete*, Spon Press, London, 1990.

Hawkins, N., Mitchell, D. "Progressive Collapse of Flat Plate Structures", *ACI Journal*, V76, pp775-808, July 1979.

Heyman, J., *Basic Structural Theory*, Cambridge University Press, Cambridge, UK, 2008.

Hurst, M.K., *Prestressed Concrete Design*, Second Edition, E&FN Spon, London, 1998.

Ibell, T.J., Burgoyne, C.J., "Behaviour of Prestressed Concrete End Blocks", *Proceedings 9th Experimental Mechanics Conference*, College Station, Texas, May 1992, pp 135-138.

Kaminetzky, D., *Design and Construction Failures*, McGraw-Hill, 1991.

Kotsovos, M.D., Pavlovic, M.N., *Ultimate Limit-state Design of Concrete Structures: a New Approach*, 1999.

Kotsovos, M.D., *Compressive Force Path Method*, Springer, London, 2014.

Lin, T.Y., Burns, N., *Design of Prestressed Concrete Structures*, Third Edition, John Wiley, New York, 1981.

MacAlevey, N., *Structural Engineering Failures: lessons for design*, Amazon, 2010.

Mosley, W. et al., *Reinforced Concrete Design*, 6th Edition, Palgrave Macmillan, 2007, UK.

Naaman, A., *Prestressed Concrete Analysis and Design-Fundamentals*, McGraw-Hill, New York, 1982.

Nilson, A., *Design of Concrete Structures*, Twelfth Edition, McGraw-Hill, New York, 1997.

Nilson, A., *Design of Prestressed Concrete*, McGraw-Hill, New York, 1978.

The Concrete Society, *Post-tensioned concrete floors: design handbook*, Technical Report TR43, Second Edition, 2005, UK.

The Concrete Society, *Movement, restraint and cracking in concrete structures*, Technical Report TR67, 2008, UK.

Thornton, C. et al., "Exposed Structure in Building Design", McGraw-Hill, New York, 1993.

Tilly, G., *Proceedings ICE Structures & Buildings*, Vol 152, Issue 1, February 2002, pp 3-16, England.

Warner, R. et al., *Concrete Structures*, Addison Wesley Longman, Australia, 1998.

Woodward, R.J., and Williams, F.W., *Collapse of Ynys-y-Gwas bridge, West Glamorgan*, Proceedings ICE, Vol. 84, Part 1, Aug 1988.

Zahn, F. Ganz, H., *Post-tensioning in Buildings*, Report 4.1, VSL International, Bern, Switzerland, 1992.

About the author:

Er. Dr. Niall MacAlevey is currently an independent consultant specializing in the analysis and design of reinforced and prestressed concrete structures, forensic engineering and the strengthening of concrete structures. He is the founder of the firm "Shamrock Consultants", and is a registered Professional Engineer in Singapore. He graduated from University College Dublin, Ireland in 1987, and completed his M.Sc. degree in "Concrete Structures" at Imperial College, London. He completed his PhD degree at the Nanyang Technological University in 1997 on "The Strengthening of Concrete Structures" and later joined the academic staff there. He obtained a PGDipTHE (Post-Graduate Diploma in Teaching in Higher Education) from the National Institute of Education in 2001. Since graduation, he has worked for a number of consulting engineering firms and specialist prestressing subcontractors in London, Cambridge, Hong Kong and Singapore.

He can be contacted at niallmacalevey@gmail.com